U0350621

北京市建筑设计研究院有限公司作品集

BIAD SELECTED WORKS 1949–2019

《北京市建筑设计研究院有限公司
作品集1949–2019》编委会 编
Edited by *Beijing Institute of Architectural Design Selected Works 1949–2019* Editorial Committee

上海 · 同济大学出版社
TONGJI UNIVERSITY PRESS

目录 CONTENTS

6 序 PREFACE

建筑中国：见证共和国走过
DESIGN IN CHINA: WITNESS TO THE DEVELOPMENT OF THE PEOPLE'S REPUBLIC OF CHINA

10 主题篇章
LANDMARK PROJECT

12 壹丨建国基业·天安门广场建筑群建设
TIAN'ANMEN SQUARE ARCHITECTURAL COMPLEX

20 贰丨开放契机·北京亚运会场馆建设
BEIJING ASIAN GAMES VILLAGE

28 叁丨入世宣言·北京CBD核心区建设
THE BEIJING CBD CORE AREA

36 肆丨联通世界·首都的机场建设
AIRPORT CONSTRUCTION IN THE CAPITAL

44 伍丨崛起之路·北京奥运建设
BEIJING OLYMPIC GAMES

54 陆丨盛世华章·主场外交建设
HOME COURT DIPLOMACY COMPLEX

62 柒丨走向世界·海外工程建设
OVERSEA CONSTRUCTION

70 捌丨和谐共生·城市环境设计
INTERNATIONAL GARDEN EXPO

78 玖丨北京故事·首都城市更新改造
CAPITAL URBAN REGENERATION

86 拾丨面向未来·设计的创新与突破
DESIGN INNOVATION AND BREAKTHROUGH

94 公共建筑
PUBLIC BUILDING

96 文化 / 体育建筑
CULTURE / SPORTS BUILDING

114 观演 / 博览建筑
PERFORMANCE / EXPO BUILDING

136 教育 / 科研建筑
EDUCATION / RESEARCH BUILDING

162 办公 / 商业建筑
OFFICE / COMMERCIAL BUILDING

196 酒店 / 会议建筑
HOTEL / CONFERENCE BUILDING

202 医疗 / 交通建筑
MEDICAL / TRANSPORTATION BUILDING

224 居住建筑
RESIDENTIAL BUILDING

236 城市设计及其他
URBAN DESIGN AND OTHERS

252 附录
APPENDIX

270 项目索引
PROJECT INDEX

序 PREFACE

建筑中国：见证共和国走过
DESIGN IN CHINA: WITNESS TO THE DEVELOPMENT OF THE PEOPLE’S REPUBLIC OF CHINA

徐全胜 Xu Quansheng

北京市建筑设计研究院有限公司党委书记、董事长
BIAD's Party Secretary & Chairman

今年是中华人民共和国成立70周年，听设计院前辈们讲，我们是1949年之后中国第一家民用建筑设计院，前身是“公营永茂建筑公司设计部”。“永茂”二字为时任北京市委书记彭真同志所题，取其“永远茂盛”之意。1949年9月，北京市人民政府派时任政府副秘书长李公侠同志筹备建立公营永茂建筑公司。在10月1日开国大典当天，“北京市公营永茂建筑公司”在当时的办公地——金城大楼楼顶垂挂了两条有公司落款的庆祝标语。同一天，4名员工以永茂建筑公司的名义参加了开国大典和晚上的提灯大游行，标志着我们与中华人民共和国同年同月同日诞生。那永存的精彩瞬间，记录下我们（现为北京市建筑设计研究院有限公司）是与共和国同龄的设计院。今天，我们纪念北京市建筑设计研究院有限公司（后文简称北京建院）成立70周年。

与共和国同龄，经历了都与城重大建设密切相关的风云际会，见证了时代潮汐下中国建筑的波澜壮阔，发展、壮大、成熟的北京建院人正用丰富的作品与先进的创作理念，用魂魄之光照亮历程，用设计文化的思想找到70年诞辰的新起点，我们的发展路标已指向百年，努力打造中国的百年企业。

70年前，北京建院迎着中华人民共和国的曙光诞生，它让我们在历史长河中寻觅中国建筑何以从自强与从容中坚守文化；70年来，北京建院阅尽风雨、内涵广博、收获颇丰，理性在召唤我们回顾并回答：中国建筑之路何以在跋涉中创作出座座高峰。面对路广而修远的新征程，我们倍感能力越强，责任越重。北京建筑经历岁月磨洗之厚重，得益于我们几代建院人始终坚守的使命、责任与价值观，得益于我们为中国建筑书写历程的事件与作品，得益于工匠的奉献精神。

我记得1999年北京建院50周年编撰学术丛书时，老院长、原城乡建设部部长叶如棠写下“堪称当代华夏第一家”的评介语；2009年北京建院60周年，我们不仅从60载时光中捡拾了60个“故事”，还串联起令北京建院人骄傲的60个“心灵地标”；从2009年的北京建院“品牌报告”到2014年“文化报告”，都彰显北京建院人爱院情怀下的宏大画卷。2019年我们再推《北京建院价值报告》，其意涵在于，立足新时代，回顾历史，展望未来，笃定、真诚、包容、善意、理性。

2018年11月22日，由北京市人民政府国有资产监督管理委员会主办、北京市建筑设计研究院有限公司联合主办，马国馨院士任总策划的“都·城：我们与这座城市”展览中展出的“首都大模型”在吸引观者目光的同时，也充分展现北京建院在北京城市建设中的贡献。在那天的中外论坛上，北京建院还代表首都建筑设计师宣读了《建筑服务社会 设计创造价值：北京建院向首都建筑设计同行及社会各界的倡议》。如果说70年我们一直在用建筑设计见证历史，那么今天推出的《北京市建筑设计研究院有限公司成立七十周年（1949—2019）院庆系列丛书》就是北京建院人在用文字的力量书写巨变，用镌刻的信仰去表述何为建筑前辈留下的真精神。作为中华人民共和国第一家民用建筑设计院，通过一个设计研究单位的工作，

This year marks the 70th anniversary of the People's Republic of China. According to seniors from our design institute, we are the first institute of civic architectural design in China after 1949, which used to be called the Design Department of the Publicly Owned Yongmao Construction Company. The Chinese characters "Yongmao", which means forever prosperous, were written by Comrade Peng Zhen, then the secretary of the Beijing Municipal Party Committee.In September 1949, the People's Government of Beijing Municipality appointed Li Gongxia, municipal deputy secretary-general, to prepare for the establishment of the Yongmao Construction Company. On October 1, the birth of the People's Republic of China, two celebratory banners were hung on the Jincheng Building, where the Yongmao Construction Company was located at the time. On that same day, four employees participated in the founding ceremony and evening lantern parade in the name of the company, marking the same birthday of the company and PRC. Those unforgettable moments tell the story that we, now named Beijing Institute of Architectural Design (BIAD) are the same age as the Republic. Today, we celebrate the 70th anniversary of the establishment of the institute.

Born on the same day as the PRC, BIAD has witnessed all the exciting advances in the urban development of Beijing as the Capital as well as a metropolitan, throughout that period. The staff members of BIAD are seeking a new start with the 70th anniversary with rich works and advanced design concepts with the light of soul illuminating our course, and the idea of design culture. We are pointing towards 100 years of development with the aim of building a 100-year-old enterprise in China.

Seventy years ago, BIAD was born in the dawn of PRC. The coincidence has since enabled us to seek the answer of why architecture in China can stick to culture by itself with self-reliance and ease in the river of history; after 70 years of development, we are called to review and answer with broad experience, comprehensive knowledge and achieved peace: how to create one accomplishment after another on the developing way of architecture in China. Faced with a long and broad new journey, we are going to take more responsibilities when growing stronger. The architecture of Beijing has over time benefitted from the work of people of BIAD, who have always stood by their missions, responsibilities and values, from events and works we wrote for the process of Chinese Architecture Era, and from the dedication spirit of artisans.

I still remember in 1999, when BIAD compiled a series of books for the 50th anniversary of the Institute, Ye Rutang, the former dean and head of the Ministry of Urban-Rural Development, commented that BIAD was "worthy to be called the leading brand in architecture design in contemporary China"; in 2009 upon the 60th anniversary, we collected 60 stories from the previous 60 years to link 60 "spiritual landmarks"; from the "Brand Report" in 2009 to the "Cultural Report" in 2014, both revealing the big picture of our people's love for BIAD. In 2019, we will publish the "Value Report" of BIAD to stand in this new era, review the history and look forward to the future with assuredness, faith, tolerance, kindness and reason.

On November 22nd, 2018, the exhibition "The Capital & The City - Beijing Be with Us" was co-hosted by the Assets Supervision and Administration Commission of the People's Government of the Beijing Municipality and BIAD, curated by scholar Ma Guoxin. The Big Capital Model attracted visitors' attention and fully showed the contribution BIAD has made to the construction of the capital. During the international forum, on behalf of architecture designers in the capital, BIAD presented a presentation entitled "Architecture Serves Society, Design Creates Value", an initiative proposed by BIAD to the Capital Architectural Design Peer in Beijing and to the Society in General. If we say we are witnessing the history of architectural design over the past 70 years, then *Beijing Institute of Architectural Design (BIAD) 70th Anniversary Collection (1949—2019)* published today is a written history of the important changes that have happened in China, expressing the true spiritual heritage engraved by our academic predecessors in the field of architecture. Through the works of a design and research institute participating in the contemporary architectural civilization of the nation, the practice of BIAD has demonstrated its ability to take part in building a Chinese architecture culture as the first civic institute of architectural design of China, a leader of advanced technology with a comprehensive strength and academic style, a practitioner of a beautiful city, a pioneer of green architecture and a demonstrator of industrial modernization. Several generations of design

参与缔造中国的建筑文化，参与写就国家的当代建筑文明，北京建院的设计实践已经说明：它是有综合实力和学术气派的先进科技的引领者、美好城市的践行者、绿色建筑的开创者和产业现代化的示范者；几代设计师们的“精神风景”为北京建院自信自强留下文化远香。

《北京市建筑设计研究院有限公司成立七十周年（1949-2019）院庆系列丛书》主要分四部：

第一：《北京市建筑设计研究院有限公司纪念集——纪事与述往》。该书以时间为轴，不是用作品，而是用典型事件与70篇文章描绘北京建院的发展演变史：北京建院何以成为中华人民共和国初创时期第一院，它从哪里来，它未来向何处去。不仅可看到令人浮想联翩的建筑艺术殿堂，如“国庆十大工程”，也有以人为本的建筑环境构成；既有首都北京70年建筑“读思录”，更有一批批建筑师、工程师的成长启示。该纪念集力求通过展示北京建院70载大事，找到北京建院为行业、为社会的无数“第一”贡献点，告诉业界北京建院的奋斗之路。在编写方式上告别传统的纪念集模式，重在讲好北京建院故事，总结北京建院精神，探索北京建院不断打造百年品牌的价值之道。

第二：《北京市建筑设计研究院有限公司作品集 1949-2019》。从天安门广场建筑群到北京城市副中心，从海南博鳌到APEC、G20峰会；从亚运会到奥运会，从园博会到世园会；从绿色城市到智慧城市，从“中国制造”到“中国创造”，可以说北京建院的历史就是中国建设事业发展的一个缩影，其技术实力、科研成果、运作经验、创新机制，对于整个中国设计行业的发展都起到了重要的推动作用。此外，1958年起，北京建院还先后承担了40多个国家的100余项援外项目，在海内外为中国建筑积累并传播巨大的影响力，这些当属早已投身的“一带一路”建设。

第三：《BIAD70周年院庆学术论文集》（涉及规划、建筑、结构、设备、电气五个分册）。它们分别围绕建筑与城市、建筑与结构、建筑与机电，从工程设计到理论实践，展现了北京建院建筑师、工程师对建筑艺术，对当今最新建筑科技诸如超大空间结构与超高层建筑，对建筑智能与健康建筑等一系列新设计理念的思考，由此丰富业界对北京建院设计思想的新认知。

第四：《北京市建筑设计研究院有限公司五十年代的“八大总”》。北京建院的沃土是如何滋养大师成长的？该书让我们从大师遗风中感悟薪火相传之力。有识之士曾说，在信念不振和乐观消逝的时代，需要学习关于故事的力量。它可打动心智、灵魂，留下难以抹灭的记忆。故事可勾勒一个个场景，还原生动的历史，它们蕴含的智慧，远胜于一个个理性的解读。据此，“生平＋评述＋作品”成为“八大总”出场的三段式。本书还将用令人信服的事例说明，什么是永不落幕的北京建院老一辈大师的学风与品格。高扬其人格精神，发扬其真思想与真性情，是传承“八大总”技艺的当代价值。

《北京市建筑设计研究院有限公司成立七十周年（1949-2019）院庆系列丛书》是一套反映北京建院人追求中国建筑技术与人文史的读本，其新意在于是集技术、文化、管理诸方面的“新记”。体现挖掘整理之“新”，反映审视与思考之“新”，更体现北京建院百年之途的品牌建设之“新”。所以，它是一套以北京建院人为根基、服务全行业的图书系；是一套具有较强可读性的技术与建筑文化兼具的图书系；更是一套对中国当代建筑史“简而有法”的图书系。

捕捉精彩，记录历史，北京市建筑设计研究院有限公司是有说服力的言者，感谢伟大祖国给予的机会。城市过去、现在和未来的发展，使得我们不仅有70载的追溯，更有百年期盼与不息的努力。向新而行的每一位北京建院人，也是与中国一起奔跑的时代创建者。我们坚信：未来建筑创作在唤起国家与城市记忆时，更要秉持“建筑服务社会、设计创造价值”的理念。我们憧憬未来，瞩目“百年北京建院”的愿景，全力打造“世界一流的建筑设计科创公司”，扎实奋进，永远在路上。

2019年10月

masters have created a spiritual and cultural atmosphere for BIAD.

Beijing Institute of Architectural Design (BIAD) 70th Anniversary Collection (1949—2019) consists of four main parts:

First: *BIAD Commemorative Biography – Documentation and History*. This volume, made up of 70 essays, uses a timeline instead of design works to describe the history of BIAD with typical events. How did BIAD become the first institute at the birth of PRC, where does it come from and where will it be in the future? You can see not only imaginative architecture designs like 10 Projects for the 10th National Anniversary, but also environmental-friendly constructions; not only a reflection of 70 years of architecture design in the capital of Beijing, but also revelations of professional maturity from groups of architects and engineers. This commemorative biography attempts to describe the contribution BIAD has made to the industry and the society by displaying the major events of the past 70 years, and by describing our struggles within the industry. Different from the traditional way of editing, this book tells the story of BIAD, sums up our working spirits and explores the way of creating a 100-year old brand.

Second: *BIAD Selected Works 1949-2019*. From Tian' anmen Square to the auxiliary city-center of Beijing, from Hainan Bo'ao to the APEC Meeting and G20 Summit; from the Asian Games to the Olympic Games, from the Horticultural Exposition to the International Garden Expo; from Green City to Smart City, and from "Made in China" to "Created in China", the history of BIAD is the epitome of national construction development. Our technical strength, research results, operational experience, and innovative mechanism have all had a great impact on the design industry of China. Beginning in 1958, BIAD also gradually completed hundreds of foreign aid projects covering more than 40 countries, accumulating and spreading the huge influence of Chinese architecture both inside and outside the country, which could be considered a precursor to the Belt and Road project.

Third: *Collected Essays in Celebrating the 70th Anniversary of the Establishment of BIAD* (consisting of five volumes: Planning, Architecture, Structures, Facilities and Electrical Engineering). These essays are separate topics on the relation between architecture and the city, structure, and electrical engineering; from engineering design to theoretical practice, showing the thinking of architects and engineers regarding the art of architecture, today's most advanced technology, such as super-large spatial structures and super high-rises, the latest design concepts such as architectural intelligence and healthy architecture, to further improve the understanding of BIAD from industry peers.

Fourth: *The Eight BIAD Masters of the 1950s*. How has the fertile soil of BIAD nourished the growth of architecture masters? A scholar once said that in an era lacking faith and optimism, we should gain the strength from stories of the past, which can touch the mind, the soul and leave us with unforgettable memories. Compared with a rational interpretation, stories can restore the vivid history by outlining scenes to reveal the wisdom of the past. Therefore, the three phases of Biography + Commentary + Work have become the path for introducing the Eight Masters. Convincing examples are quoted in the book to elaborate on the eternal academic style and characteristics of the older masters of BIAD. It is the contemporary value of inheriting craftsmanship from the Eight Masters to raise their spiritual personality and develop their true thoughts.

Beijing Institute of Architectural Design (BIAD) 70th Anniversary Collection (1949—2019) is a set of readings that reflect upon the chase of BIAD of Chinese architectural techniques and cultural history, and the new ideas that lay in the combined new record of technology, culture and management. The "new" is also reflected in digging and sorting, examining and thinking, and more in the construction of building up a BIAD brand with a history of 100 years. It is a set of books based on BIAD staff who served the whole industry, highly readable in techniques and the culture of architecture as well as a simple but regulated contemporary architectural history of China.

Thanks to the opportunities provided by our great motherland, the BIAD has become a persuasive storyteller because it has captured wonders and documented the history. The past, present and future development of the city not only gives us 70 years of retrospection, but also gives us expectations to the arrival of the centennial with unremitting efforts. As a BIAD member heading into a new future, and one who is also a founder in the new era with China, we firmly believe that future architectural creation should adhere to the concept of "architecture serves the society, design creates value" when evoking national and urban memories. We look forward to the future with a focus on the vision of 100 years of BIAD building a "architecture design company with a world-class capability of technology innovation". BIAD is always on the road of innovation, dedication and learning.

October, 2019

主题篇章
LANDMARK PROJECT

12 壹 | 建国基业 · 天安门广场建筑群建设
TIAN'ANMEN SQUARE ARCHITECTURAL COMPLEX

20 贰 | 开放契机 · 北京亚运会场馆建设
BEIJING ASIAN GAMES VILLAGE

28 叁 | 入世宣言 · 北京CBD核心区建设
THE BEIJING CBD CORE AREA

36 肆 | 联通世界 · 首都的机场建设
AIRPORT CONSTRUCTION IN THE CAPITAL

44 伍 | 崛起之路 · 北京奥运建设
BEIJING OLYMPIC GAMES

54 陆 | 盛世华章 · 主场外交建设
HOME COURT DIPLOMACY COMPLEX

62 柒 | 走向世界 · 海外工程建设
OVERSEA CONSTRUCTION

70 捌 | 和谐共生 · 城市环境设计
INTERNATIONAL GARDEN EXPO

78 玖 | 北京故事 · 首都城市更新改造
CAPITAL URBAN REGENERATION

86 拾 | 面向未来 · 设计的创新与突破
DESIGN INNOVATION AND BREAKTHROUGH

壹

建国基业·天安门广场建筑群建设
TIAN'ANMEN SQUARE ARCHITECTURAL COMPLEX

伟大开篇，铸就城市经典
GREAT BEGINNING, FOUNDING OF A CITY CLASSIC

70年前，1949年10月1日下午三点，天安门广场，毛泽东主席用他那带着湖南口音的洪亮声音，向全世界庄严宣告：中华人民共和国中央人民政府成立了！这一庄严宣布开辟了中国历史新纪元。开国大典的举办也让天安门广场变成了中华人民共和国的政治中心，变成了中国的象征。

在新中国诞生的同时，北京建院（当时名称为北京市公营永茂建筑公司设计部）同时诞生。作为一家与共和国同龄的大型国有建筑设计咨询机构，北京建院是新中国第一家建筑设计院，像一粒种子似的，承担起首都的建设任务。

从建国初的天安门城楼（落架修缮）、观礼台建设，到以人民大会堂、中国革命历史博物馆为代表的1959年"国庆十周年"十大工程，再到毛主席纪念堂、国旗基座、广场改造、全国人大常委会办公楼、全国人大常委会会议厅改扩建等一系列工程都位于天安门广场这片神圣的区域。北京建院为这片共和国的心脏地带贡献了自己的智慧和心血，正如北京建院党委书记、董事长徐全胜所说，"北京建院在积极投身于城市建设的同时见证着首都北京的变迁"。

天安门广场的建设不仅在中国的发展进程中意义重大，同时也具有极高的历史和艺术价值。2007年，人民大会堂入选《北京优秀近现代建筑保护名录（第一批）》；2011年，天安门观礼台荣获北京国际设计周"设计北京大奖"；2013年，天安门广场建筑群获菲迪克百年重大建筑项目优秀奖。

It was seventy years ago, at three o'clock in the afternoon on October 1, 1949, at Tian' anmen Square, that Chairman Mao Zedong made a solemn announcement to the whole world with a resonant voice: The State Council of the People's Republic of China is established today! This was opening up a new era in the history of China. After the founding ceremony, Tian' anmen Square then became the political center of the People's Republic of China, the symbol of China.

BIAD was founded along with the birth of PRC. As the first large-scale state-owned architectural design and consulting institute, BIAD shouldered the responsibility of the construction work in the capital.

At this sacred place of Tian' anmen Square, BIAD was devoting all of its wisdom and efforts to this piece of heartland of the PRC, including a redevelopment project of the Tian'anmen Rostrum and construction of a viewing platform.

The 10 Projects marking the 10th national anniversary in 1959, includes the Great Hall of the People, the Chinese Revolution Museum and Chinese History Museum, Chairman Mao Memorial Hall, the base for the national flag, the renovation of the square, establishment of office building and extension and renovation of the Conference Hall of the Standing Committee of National People's Congress of the Republic etc. Just as Mr. Xu Quansheng, BIAD's chairman and Party secretary, said: "BIAD is actively participating in the city construction and witnessing the changes of Capital Beijing."

The construction of Tian' anmen Square not only marks a great milestone in the development of China, but also has a high historical and artistic value. In 2007, the Great Hall of the People was listed into the "Conservation List of Beijing Outstanding Modern and Contemporary Architecture (First Group)"; in 2011, the viewing platform of Tian' anmen Square was awarded the "Beijing Design Award" during the Beijing Design Week; in 2013, the Tian' anmen Square Architectural Complex was given an award for merit by the FIDIC Centennial Awards.

天安门建筑群

Tian'anmen Square

天安门城楼重建工程 / 天安门观礼台

The Restoration Project of the Tian'anmen Square Rostrum / Tian'anmen Square Viewing Platform

天安门城楼重建工程

Restoration Project of the Tian'anmen Square Rostrum

建成时间 Completion Year：1970

1969 －1970 年的天安门城楼重建工程，由周恩来总理担任总指挥，是建国 20 年来最大的古建筑项目，也是清代晚期以来少有的大型古建筑修建工程。尤其天安门城楼是载入国徽的，显得更加神圣。

In 1969—1970, the restoration project of the Tian' anmen Square Rostrum, led by Premier Zhou Enlai, was the biggest historical architecture project in the first 20 years after the birth of PRC, and also a rare large-scale restoration project since the late Qing Dynasty. It became more sacred when the rostrum was crowned with the National Emblem.

天安门观礼台

Tian'anmen Square Viewing Platform

建成时间 Completion Year：1954

建设面积 Building Area：4 008m²

天安门观礼台位于天安门前方东西两侧，看台平缓的坡度成功地弱化了它巨大的体积，为拥有 500 多年历史的天安门城楼营造了更加恢弘的气势。中国的建筑学界曾经评论："这观礼台好就好在让人感觉不到它的存在。"

The Tian' anmen Square Viewing Platform is located on both the east and west sides in front of the square, with a gentle slope to deliberately diminish its enormous size but increase the magnificence of the Tian' anmen Square Rostrum which has a history of more than 500 years. The academic community of Chinese architecture studies once commented: "The advantage of this viewing platform is that it doesn't make people feel its existence."

毛主席纪念堂
Chairman Mao Memorial Hall

建成时间 Completion Year：1977

建设面积 Building Area：28 225m^2

毛主席纪念堂是天安门广场南侧的核心建筑。平面布局严整，造型简洁，给人以稳重平衡的感觉，具有强烈的中心感和庄严肃穆的艺术效果。建筑从开始设计到土建竣工仅用了6个月的时间，创造了我国建筑史上的新纪录。

The Chairman Mao Memorial Hall is the core piece of architecture at the south end of Tian' anmen Square. With a clear layout and simple form, the hall gives a modest but balanced feeling to people, and a strong sense of center with a solemn artistic effect. The construction took only six months from the beginning of design to the completion of construction, creating a new record in the history of Chinese architecture.

人民大会堂
Great Hall of the People

建成时间 Completion Year :1959

建设面积 Building Area :171 800m^2

人民大会堂是1959年“国庆十周年”十大工程之一，从1958年9月开始征集建筑方案，仅用了10个多月的时间，其精美程度，不仅远远超过我国当时原有同类建筑水平，在世界上也属一流。

The Great Hall of the People is one of the 10 projects of the 10th national anniversary. It took less than 11 months from the submission of the architectural designs in September 1958, to the completion in August 1959. However, its exquisiteness not only far exceeded the construction level in our country at that time for similar architecture but was even first class in the world in the 1950s.

中国革命博物馆与中国历史博物馆
Chinese Revolution Museum and Chinese History Museum

建成时间 Completion Year：1959

建设面积 Building Area：65 152m²

中国革命博物馆与中国历史博物馆是1959年"国庆十周年"十大工程之一，是中国建立的第一座国家博物馆。建筑坐落在一个宽大的基座上，设计采用院落式建筑布局，革命、历史两个博物馆分别位于南北两个院落，中间前部的院子有空廊通向广场，同时与南北两个院子相连贯。

The Chinese Revolution Museum and Chinese History Museum is one of the top 10 projects built for the 10th national anniversary in 1959, and the first national museum built in China. The structure sits on a large base and adopts the layout of courtyard-style architecture. The two museums of revolution and history are respectively located in the north and south courtyards, which are connected by an open corridor leading from the front courtyard to the square in the middle.

人民大会堂

Great Hall of the People

贰

开放契机·北京亚运会场馆建设
BEIJING ASIAN GAMES VILLAGE

完善现代化，亚洲雄风震天吼
THE CONSUMMATION OF CHINA' S MODERNIZATION

对于中国人来说，1990年北京亚运会是永远抹不去的记忆。这场赛事是中国1978年改革开放之后第一次举办国际性体育赛事，是中国对外开放、北京走向国际化的重要时间点。当年为举办亚运会，举国上下出力、出钱，一批现代化的亚运会基础设施拔地而起。很多人为北京新地标的宏伟吃惊，认为至少超越时代10~15年。

为办好这次亚洲盛会，中国做了大量的准备工作，北京兴建了以亚运会主体育场为主体的奥林匹克体育中心和亚运村，并建设了大量的立交桥和宽敞的马路，城市面貌焕然一新。在这场火热的城市建设大潮中，北京建院历经7年时间，承担了包括国家奥林匹克体育中心、亚运村在内的80%的设计任务。

从参加国际大赛到举办国际大赛，中国希望借此对外扩大开放，在国际上重新树立大国形象。而在城市建设上，则以此为“抓手”加快城市建设，提高国人物质文明和精神文明，北京亚运会成了刚开始改革开放的中国向世界展示民族文化、精神气质和大国形象的重要舞台。之后，中国社会发生了翻天覆地的变化，并成为世界第二大经济体，虽然期间发生了无数的大事，但是1990年的亚运会始终是中国人心中最美好的记忆。

The 1990 Beijing Asian Games is an unforgettable memory for the Chinese people. It was the first international sports event held by China after the Reform and Opening began in 1978, which marked the important moment when China opened up and Beijing went global. To host the Asian Games, a group of modern infrastructures were constructed up with the whole nation's effort and money. These astonishing new landmarks in Beijing were believed to be at least 10 to 15 years ahead of their time.

China made a lot of preparations for the great Asian Games, including the National Olympic Sports Center, where the main sports stadium and the Asian Games Village were built in Beijing, as well as a lot of overpasses and widening of roads. The city of Beijing was given a brand-new look. During this boom in city construction, BIAD spent seven years to finish 80 percent of the design tasks, including the Asian Games Village and the National Olympic Sports Center.

From participating in such events to the actual holding of these international games, China made an effort to open up and establish an image of a great country to the world. This was also an opportunity to speed up city construction and develop culture and spiritual civilization. The Beijing Asian Games became the main stage for China to show its national culture, spirit, and character to the world after the period of Reform and Opening. Since then, Chinese society changed dramatically, and China became the world's second largest economy. The 1990 Beijing Asian Games remains one of the best memories in the minds of the Chinese people, among other big events.

国家奥林匹克体育中心
National Olympic Sports Center

国家奥林匹克体育中心 一期
National Olympic Sports Center Phase I

建成时间 Completion Year : 1989

建设面积 Construction Area : 103 270m²

国家奥林匹克体育中心位于北京市北四环中路南侧，与奥林匹克公园、奥林匹克森林公园和奥运村隔路相望，是集竞赛训练、全民健身、休闲娱乐为一体的体育基地、体育公园。主要项目有6000座的体育馆和游泳馆，20 000座的田径比赛场和练习馆，2000座的曲棍球场，以及田径和足球练习场、投掷场、垒球场、网球场、体育博物馆、武术研究院、医务测试及附属用房等。

The National Olympic Sports Center is located on the south side of the North Fourth Ring Road in Beijing, while the Olympic Park, Olympic Forest Park and Olympic Village are on the other side of the road. It is an integrated sports base and park for competition training, national fitness and recreation. Featured projects are the stadium and natatorium with 6,000 seats, athletics field and arena with 20,000 seats, hockey rink with 2,000 seats, athletics and football arena, field for throwing events, softball field, tennis court, Sports Museum, Research Institute of Martial Arts, medical center and other ancillary buildings.

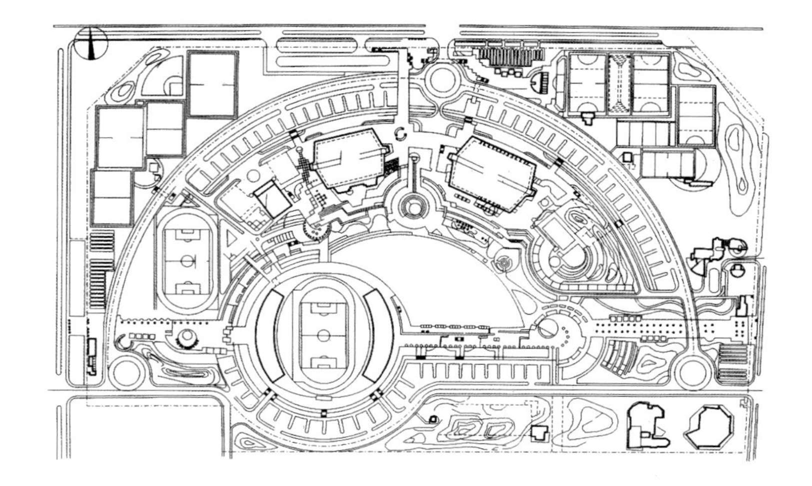

北京工人体育场
Beijing Workers' Stadium

建成时间 Completion Year :1959
建设面积 Building Area :87 080 m^2

北京工人体育场是1959年“国庆十周年”十大工程之一，始建于1959年，1990年亚运会前及2008年奥运会前，工人体育场的建筑及配套设施分别进行了升级、改造、加固。竞赛场为椭圆形，场内设置有400m跑道、足球场、跳远及助跑道、跳高、铅球、链球、铁饼、标枪、手榴弹等各项场地，实际可容观众78 000多人。1990年，工人体育场作为北京亚运会开幕式、闭幕式的场地，完整见证了这一重要历史时刻。

The Beijing Workers' Stadium is one of the 10 projects for 10th national anniversary, first built in 1959. Before the 1990 Beijing Asian Games and the 2008 Beijing Olympic Games, the building and supporting facilities of the Workers' Stadium had been upgraded, transformed and reinforced separately. The competition venue is oval, with a 400m-long track, football field and various fields for long jump and auxiliary track, high jump, shot put, hammer throw, discus throw, javelin throw, grenade and other activities, which can actually hold more than 78,000 spectators. In 1990, the Workers' Stadium was the venue for the Opening and Closing Ceremony of the Beijing Asian Games, which witnessed this important moment in history.

第十一届亚运会运动员村及配套设施
The 11th Asian Games Village and Supporting Facilities

建成时间 Completion Year :1985

建设面积 Building Area :520 000m^2

亚运村紧靠奥运中心，各类建筑配套功能和规模的设计均按双重使用考虑——既为运动会服务，又考虑会后长期的使用，如旅游商贸、会展等。以新闻中心为例，其为具有多种使用功能的国际会议大厦。

The Asian Games Village sits beside the Olympic Center, whose design and supporting facilities were built for dual use — not only for athletic events, but also for long-term use after the games, such as tourism, trade and conventions. For example, the News Center is an international conference building with multiple functions.

国家奥林匹克体育中心

National Olympic Sports Center

入世宣言·北京CBD核心区建设
THE BEIJING CBD CORE AREA

面向未来的城市核心
A CITY CENTER FACING THE FUTURE

2001年11月10日，在卡塔尔多哈举行的世界贸易组织（WTO）第四届部长级会议通过了中国加入世界贸易组织法律文件，标志着中国终于成为世界贸易组织新成员。北京作为中国的首都和面向世界的窗口，其城市建设也进入了全新时期。集约型发展、高密度的紧凑型城市成为这一阶段北京城市发展的全新写照。城市期待更高的发展理念，也期待更加整合的土地混合使用模式和效率。北京CBD核心区正是在这样的背景下诞生的。

在21世纪的今天，城市的功能空间也越来越复合，而这也让未来城市的构想在今天的超级建筑群中逐步得以实现。作为牵头整合北京CBD核心区建设资源的北京建院不仅关注如何克服安全与建构上的困难，更开始重视新能源和可持续发展的技术体系在超高层建筑中应用的探索。从新的城市营造理念入手，高效地利用土地资源，用精致的发展方式代替无序蔓延，让城市在功能、规模和结构上更加紧凑与高效：从整体到局部，从地上到地下，从规划整合到专项设计……城市空间的一体化在这里达到了极致：集约性的规划景观、商业服务、综合避险、市政交通、集中的能源供给以及无处不在的智慧城市体系。

On November 10, 2001, the Fourth Ministerial Conference of the World Trade Organization (WTO) held in Doha, Qatar, approved of the legal documents for China's entry into the WTO, marking the final success of China in becoming a new member of the organization. As the capital of China and its window to the world, Beijing also underwent a new period of urban development, when consolidation and high-density that were characteristics of a compact city became the new profile of its urban development. CBD was then born in expectation of a better development concept, with more integrated mixed land use and higher efficiency.

Today in the 21st Century, the conception of a future city is gradually being realized in these super architecture complexes where urban spaces are becoming more multi-functional. Leading the integration of the Beijing CBD construction resources, BIAD worked on the problems of security and composition, and began to develop the application of new energy resources and sustainable techniques in the construction of super high-rise buildings. We started from new concepts of city building: use land resources with high-efficiency, replace disordered sprawl with delicate development methods, making the city more compact and efficient in terms of function, scale and structure from the whole to the part, from above ground to underground, and from overall planning to detailed design. The integration of urban spaces is now reaching an extreme: compact landscape planning, business services, a comprehensive way of hedging, integrated municipal traffic, central resource supply and a smart city system everywhere.

北京CBD

Beijing CBD

中信大厦
CITIC Tower

建成时间 Completion Year : 2019
建设面积 Building Area : 437 000m²
合作公司 Partner : KPF/ 中信建筑设计研究总院有限公司 CITIC General Institute of Architectural Design and Research Co.,Ltd / Arup/ PBA 柏诚 /TFP

中信大厦是首都的新地标，与国贸建筑群、中央电视台和银泰中心等构成了北京新的天际线。项目在1.15公顷的基地面积上建设了地上35万m²、地下8.7万m²的建筑规模，建筑总高度528m，地上112层，地下8层，深度40m，基础桩深100m，容积率大于30。它既是北京最高和最深的建筑，同时技术的复杂性、系统的多样性、建筑的超高尺度、建造工期、成本的因素也对设计与建造提出空前的技术挑战。

CITIC Tower is the new landmark of the capital, forming the new skyline of Beijing together with the World Towers Complex, CCTV Headquarters and the Yintai Center. On a site of 1.15 hectares in area, this project has a building scale of 350,000 square meters above the ground and 87,000 square meters below ground, with a total height of 528 meters, 112 floors above ground and 8 floors underground, a depth of 40m, a foundation pile depth of 100m and a plot ratio of more than 30. It is both the tallest and deepest building in Beijing, and the technical challenges of design and construction are unprecedented due to the complexity of techniques, diversity of systems, ultra-high scale of buildings, construction duration and cost.

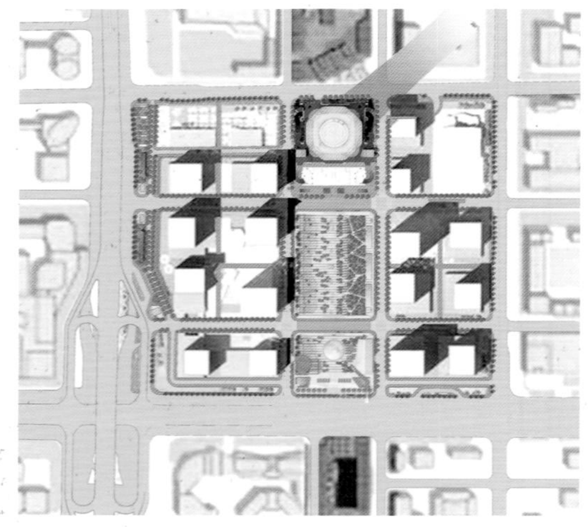

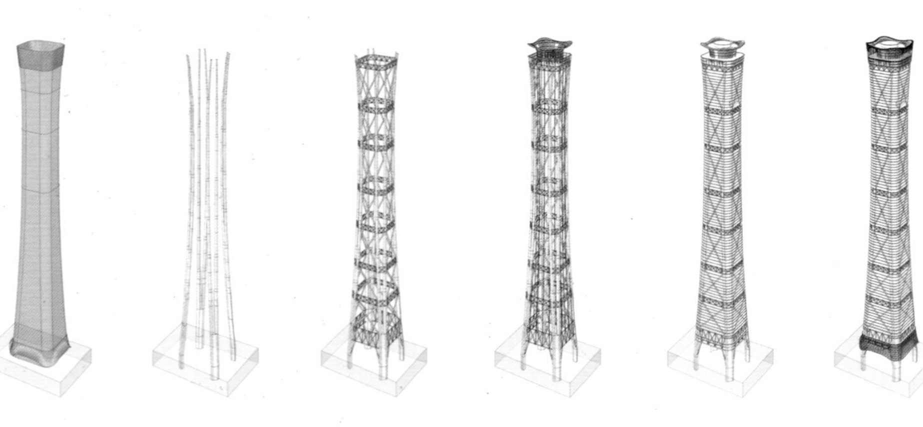

北京CBD核心区地下公共空间市政交通基础设施及配套工程

Municipal Transportation Infrastructure Project in Underground Public Space of the Beijing CBD Core Area

设计时间 Design Began：2015

建设面积 Construction Area :507 568m²

北京CBD核心区地下公共空间建设面积占CBD核心区总建设面积的1/3。作为系统的地下空间建设项目，集交通、市政、景观以及综合防灾等多种功能于一体。

The construction area of the underground public space of the Beijing CBD core area accounts for one-third of the total construction area. As a systematic underground space construction project, it integrated multiple functions such as transportation, municipal administration, landscaping and comprehensive disaster prevention.

中国人寿金融中心（北京）

China Life Financial Center (Beijing)

建成时间 Completion Year：2019

建设面积 Building Area：162 000m²

合作公司 Partner：SOM建筑设计事务所 Skidmore, Owings & Merrill LLP

中国人寿金融中心以190m的板式超高层与周边建筑形成了完整的CBD核心区的城市节点，建筑底部退让的半室外通高廊架可作为小区域内的城市客厅。该建筑是北京建院在CBD核心区落成并正式启用的第一栋建筑，也是中国第一个LEED、WELL双金级认证的大厦，本项目已经成为CBD核心区整体形象的重要组成部分。

The China Life Financial Center, a 190m-high slab-type super high-rise, forms a complete urban node of the CBD core area with surrounding buildings. The semi-outdoor double-height open space at the bottom of the building can be seen as the urban living room in a small scale. It is the first building designed by BIAD that has been completed and officially opened in the CBD core area, and also the first LEED and WELL double-gold certified building in China. This project has become an important component of the overall image of the CBD core area.

北京 CBD
Beijing CBD

肆

联通世界·首都的机场建设
AIRPORT CONSTRUCTION IN THE CAPITAL

跨越式发展的中国航空
LEAP-FROG DEVELOPMENT OF AVIATION INDUSTRY IN CHINA

从1958年中国第一个建成投入使用的大型民用机场，到中国第一个卫星式、具有现代意义的国际航空港，再到世界最大的航站楼；从初步开启拓展以机场为核心的经济开发模式，到超大型国际航空综合交通枢纽；从自力更生、自主设计，到与国际接轨、合作设计，再到最终实现了创作理念和技术革命的引领……

经过半个多世纪的探索和实践，北京建院见证了首都民航事业的跨越式发展。除首都机场以外，近十年来，北京建院先后承担了昆明长水国际机场、深圳宝安机场T3航站楼、南宁吴圩机场、桂林机场T2 航站楼、长春机场T2航站楼、海口美兰国际机场T2航站楼等大型、超大型机场航站楼的工程设计。

从首都机场T1、T2航站楼，到T3航站楼，再到北京大兴国际机场，北京建院在实践中不断刷新航站楼设计领域的技术创新和突破，这场空中国门的巨大变革正是中国伟大开放和飞速发展的缩影。作为空中国门巨大变革的亲历者，北京建院始终为中国的机场建设不间断地提供设计服务，逐渐形成机场航站区规划、陆侧交通设计、航站楼建筑设计等方面的核心设计理念和方法，实现了创作理念和技术革命的引领。

Over the past 70 years, China has made great strides in the field of architecture: from its first large civic airport, which went into operation in 1958, to the country's first modern international airport with satellite design, to the world's biggest terminal; from the preliminary attempt to explore an airport-centered economic development model, to a super-large comprehensive transportation hub for international transportation; from self-reliant and independent designs, and working according to international standards and with collaborative desgigns, and to finally leading the trend of design and technical revolution.

Through half a century of exploration and practice, BIAD has witnessed the rapid development of the capital's civic aviation industry. In addition to Capital Airport, BIAD has also designed large or super-large airport terminal designs, such as the Kunming Changshui International Airport, Terminal 3 of the Shenzhen Bao'an International Airport, Nanning Wuxu International Airport, Terminal 2 of the Guilin Liangjiang International Airport, Terminal 2 of the Changchun Longjia International Airport and Terminal 2 of the Haikou Meilan International Airport, one after one in recent 10 years.

From Terminal 1 and 2 of Beijing Capital International Airport to Terminal 3, and to Beijing Daxing International Airport, BIAD is constantly making technical innovations and breakthroughs through practice in the field of terminal design. This huge revolution of national gateway in the air is the epitome of China's Reform and Opening and rapid development. As a witness of the revolution in civic aviation, BIAD has consistently provided design services for China's airport construction and has gradually formed core design concepts and methods in terms of airport regional planning, transportation design on the ground, terminal architecture design etc., which gave it a leading position in design trends and technical revolution.

首都国际机场 T3 航站楼

Terminal 3 of Beijing Capital International Airport

首都国际机场 T1 航站楼

Terminal 1 of Beijing Capital International Airport

建成时间 Completion Year :1979

建设面积 Building Area :58 000m^2

1980年1月1日，首都机场T1航站楼及停机坪、楼前停车场等配套工程建成并正式投入使用，这是我国当时最先进的机场。此后T1航站区经历了前后7次改扩建，包括行李分拣大厅的改建，主楼及引桥等的扩建，并将航站楼面积增至8万m^2，旅客吞吐量提高到每年800万人次。

On January 1, 1980, Terminal 1 of Beijing Capital International Airport and its airplane apron, building parking lot and other supporting projects were completed and put into use. This was the most advanced airport in China at that time. After that, Terminal 1 has gone through seven renovations and expansions, including the luggage sorting hall, the main building and the approach bridge. The area of terminal has been expanded to 80,000 square meters, and the handling capacity of passengers has been increased to accommodate eight million passengers per year.

首都国际机场 T2 航站楼

Terminal 2 of Beijing Capital International Airport

建成时间 Completion Year: 1999

建设面积 Building Area :327 000m^2

1999年11月1日，首都机场T2航站楼投入使用。工程从1995年开始动工建设，历经4年建成。包括一座32.7万m^2的国内单体建设面积最大的航站楼，以及建设面积16.7万m^2、可停放5000多辆汽车、当时亚洲最大的停车楼等17个配套项目。

On November 1, 1999, Terminal 2 of Beijing Capital International Airport was put into use. The project started construction in 1995 and was completed in 4 years. It included the largest terminal building in China with 327,000 square meters in area, and the largest parking building in Asia at that time with 167,000 square meters in area, and the capacity to park more than 5,000 cars. In addition, there are 17 other supporting projects.

首都国际机场T3航站楼

Terminal 3 of Beijing Capital International Airport

建成时间 Completion Year : 2008

建设面积 Building Area : 902 000m^2

合作公司 Partner : NACO / Foster+Partners / ARUP

北京首都国际机场T3航站楼是中国第一个真正意义上的枢纽机场航站楼，由主楼、国际候机廊和楼前交通系统组成。南北向长2900m，是当时世界上最大的单体建筑，整个工期仅为三年零九个月。T3航站楼投入使用后，首都机场滑行道将由71条增为137条，停机位由164个增为314个，第三条跑道可供世界最大的空中客车A380客机起降。

Terminal 3 of Beijing Capital International Airport is the first hub terminal in China, consisting of a main passenger terminal, waiting hall for international departures and a transportation center. It was the largest single building in the world at the time of completion with a length of 2,900m running in the direction of north to south. The construction only took three years and nine months. After the opening of Terminal 3, the number of taxiways increased from 71 to 137, and the number of parking spaces for planes increased from 164 to 314. The third taxiway will make it possible for the world's largest commercial airliner, the Airbus A380, to take off and land at the airport.

北京大兴国际机场
Beijing Daxing International Airport

建成时间 Completion Year :2019

航站区建设面积 Terminal Construction Area : 1.41km²

合作公司 Partner : 中国民航机场建设集团公司

北京大兴国际机场是北京在21世纪面向未来发展的超级工程，满足首期4500万人次旅客吞吐的设计容量。航站楼由中央主楼和5条互呈60°夹角的放射状指廊构成，航站楼以北的综合服务楼平面形状与航站楼的指廊相同，与航站楼共同形成了外包直径1200m的总体构型。机场中心区域的支撑间距达200m，所形成的无柱空间可以完整地放下一个水立方。

Beijing Daxing International Airport is a super project of Beijing in 21st Century to meet the needs of future development, designed with an annual handling capacity of 45 million passengers. The terminal is composed of a central main building and five airside concourses at an angle of 60 degrees to each other. The comprehensive service building at the north of the terminal has the same shape as the concourses in the terminal and forms an overall configuration with the terminal building in an outer diameter of 1,200 meters. The distance between supporting columns in the central area of the airport is up to 200 meters, which forms a column-free space large enough for a complete Water Cube.

践行新发展理念 打造精品工程 样板工程 平安工程 廉洁工程
不忘初心 牢记使命

北京大兴国际机场

Beijing Daxing International Airport

伍

崛起之路·北京奥运建设
BEIJING OLYMPIC GAMES

助力双奥城 圆梦冬奥会
TWO OLYMPIC GAMES IN ONE CITY

百年奥运，中华圆梦。2001年7月13日，国际奥委会主席萨马兰奇宣布，中国北京凭借其过人的优势和完美的陈述报告，在5个2008年夏奥会申办城市中脱颖而出，夺得2008年夏奥会举办权。

1991年，中国首次申办奥运会，却以一票惜败。2000年，北京再次启动申办奥运会的工作，北京建院承担了申奥场馆的全部设计任务，按照国际奥委会的标准，对30多个场馆进行设计规划。经过一年半的规划、论证、设计和数十遍的修改，在北京奥申委代表团向国际奥委会送交的《申办报告》中，北京建院完成的奥运场馆及奥林匹克公园的设计方案，获得了国际奥委会委员们的好评，为申奥成功做出了突出贡献，更为中国体育建筑奠定了坚实的基础。

2008年，奥林匹克精神在华夏大地上传播，中国的国家形象被重新打造，人们的国家自豪感达到了一个前所未有的高度。在北京夏奥会的场馆建设中，北京建院承担了35项奥运场馆及配套工程的设计工作，共完成了270万m²的设计任务，占全部场馆总面积的40%。

而在即将到来的2022年北京冬奥会、冬残奥会的筹办和建设中，北京建院承担了场馆的新建、改扩建设计项目10项，以及前期咨询、服务项目8项，严格按照国际奥委会的标准、冬奥组委和市重大项目办的要求，统筹制定统一的技术和设计标准，以打造“北京冬奥工程质量”金名片为目标，统筹协调推进，确保所有项目高标准、高质量地完成。

With the awarding of the 2008 Summer Olympic Games, China's dream came true. On July 13, 2001, International Olympic Committee (IOC) President Juan Antonio Samaranch announced that among five bidding cities Beijing won the majority of votes and so was awarded the right to host the 2008 Summer Olympic Games as a result of its superior advantages and perfect presentation.

In 1991, China narrowly lost its first bid to host the Olympic Games by just one vote. In 2002, Beijing restarted the preparation work for a new bid, and BIAD was assigned to do all the design and planning work for more than 30 stadiums for the Olympic Games, according to the standards of the International Olympic Committee. After one and a half years of planning, demonstration, design and revising dozens of times, BIAD finished the design of the Olympic stadiums and Olympic Park, which received positive reviews from IOC members as a winning part in the Bidding Report submitted by the Beijing Olympic Games Bidding Committee, contributed to the success of winning the bid and laid a solid foundation for the future design and planning of sports venues in China.

In 2008, the Olympic spirit spread all over China, rebuilding the national image of China and unprecedentedly enhancing our national pride. Among all the construction work for stadiums for the Beijing Summer Olympic Games, BIAD was awarded 35 design projects for stadiums and supporting facilities, covering 2700,000 m2, or 40 percent of the total area.

In preparation and construction for the coming 2022 Beijing Winter Olympic Games and Winter Paralympic Games, BIAD is responsible for 10 projects that will be newly-built or renovated and eight projects that will require preliminary consulting services. Strictly in accordance with IOC standards, the requirements of the Beijing Organising Committee for the 2022 Olympic and Paralympic Winter Games and Beijing City Major Projects Construction Headquarters Office, BIAD comprehensively set unified technical standards and design requirements, aiming to ensure a high level of quality for all the construction work.

北京奥林匹克建筑群

Beijing Olympic Venue

北京奥林匹克公园中心区景观设计

The Landscape Design of the Beijing Olympic Park Central District

建成时间 Completion Year：2008
建设面积 Construction Area：11 122 488m^2
合作公司 Partner：北京市市政工程设计研究总院有限责任公司 BMEDI / 北京水利规划设计研究院 BIW / 北京中国风景园林规划设计研究中心 China Research of Landscape Architectural Design and Planning / 北京城建设计发展集团股份有限公司 UCD / 清华大学建筑设计研究院有限公司 THAD / 中国建筑设计院有限公司 CADG / 北京市文物建筑保护设计所 Beijing Heritage and Relics Preservation Design Institute / 北京齐欣建筑设计咨询有限公司 Beijing Qixin Architectural Design Consulting Co.,Ltd

奥林匹克公园中心区的景观设计延续了北京城市的棋盘格网布局，设计风格简约、现代、宏大，三条相互渗透的轴线和一座下沉花园成为设计的最大特征。三条轴线分别是体现庄重理性的中轴、体现人文自然的绿轴、体现生态科学的水轴，在一个相对紧密的空间内相互联系、交融，形成统一整体。设计将城市空间作为整体考虑，将"科技、人文、绿色"的理念融入其中，形成一个大景观空间的概念，是中国传统文化与现代景观设计手法的充分结合。北京奥林匹克公园的建设使北京城市的古老轴线向北延伸并使其更为丰富，翻开了北京城市建设的新篇章。

The landscape design of the Olympic Park Central District was modeled on the layout of the chessboard-grid form in Beijing. "Simple, modern, and grand" correctly describe this style. Three axes which are intersected, and a sunken garden form the main features of the design. The three axes are the central axis, which symbolizes solemn rationality, the green axis which symbolizes nature, and the water axis, which symbolizes ecological science. All these three axises connect, communicate, and are integrated with each other, becoming a unified whole. The design puts into consideration the urban space as a whole, integrating the concepts of "science, technology, and green" to form a larger landscape space. It is the embodiment of the combination of Chinese traditional culture and modern landscape design. The construction of the Beijing Olympic Park extends and enriches the ancient south-north axis further up north, opening a new chapter in the architectural history of Beijing.

国家体育馆

National Stadium

建成时间 Completion Year：2008

建设面积 Building Area：80 890 m^2

合作公司 Partner：德国慕尼黑集团 Munich Group / 北京城建设计研究总院 Beijing Urban Engineering Design & Research Institute

国家体育馆位于北京奥林匹克公园中心区的南部，以中国“折扇”为设计灵感，是国内最大的双向张弦钢屋架结构体系，采取由南向北的波浪式造型，屋面轻盈而富于动感。国家体育馆是北京夏奥会三大主场馆之一，在2008年北京夏奥会中，国家体育馆先后承担了体操、蹦床、手球项目的比赛。该馆经改造，在2022年冬奥会中将作为男子冰球比赛场馆。

The National Stadium is located in the south of the Beijing Olympic Park Central District. The design of the stadium was inspired by the Chinese folding fan. The building has the largest scale two-way string steel truss structure system in China. Adopting the waving shape from south to north, the roof gives people the feeling of lightness and action. The National Stadium was one of the three main venues of the Beijing Summer Olympics, which held competitions in gymnastics, trampolining and handball. This venue will be renovated for the 2022 Winter Olympics and will be the venue for men's ice hockey competition.

五棵松体育馆
Wukesong Stadium

建成时间 Completion Year：2008

建设面积 Building Area：63 000m^2

五棵松体育馆建筑为一个金色方形体，其外立面由一块块的铝合金板围合而成。建筑巧妙地利用先进的分层进出场馆的方式，实现了观众与运动员、官员和管理人员的分流。同时在“节俭办奥运”原则的指导下，设计优化后的五棵松体育馆造价大幅下降，而且施工难度降低，安全性能提升。五棵松体育馆曾为2008年北京夏奥会篮球比赛场地，在2022年北京冬奥会中将作为女子冰球比赛场馆。

Wukesong Stadium is a golden square cube, with the façade enclosed by pieces of aluminum plates. The design smartly divides the flow of movement among audience, athletes, officials and management personnel by adopting separate levels for entering and exiting, the attendees enter the competition hall from the platform outside. At the same time, under the principle of “economical operation for Olympics”, the cost of Wukesong Stadium was greatly reduced by optimized the design, reducing the difficulty of construction, and enhancing the safety of the structure. The Wukesong Stadium was the basketball venue of the 2008 Beijing Summer Olympic Games, and it will be the venue for women's ice hockey at the 2022 Beijing Winter Olympic Games.

国家会议中心

National Conference Center

建成时间 Completion Year :2008

建设面积 Building Area :269 991m²

合作公司 Partner :RMJM

国家会议中心位于鸟巢和水立方之北，2008年夏奥会期间作为击剑馆、国际广播中心使用。建筑外形优美，立面设计取意中国古代建筑屋檐的曲线，将传统的建筑形式进行现代的演绎，同时又象征一座桥梁，与奥运公园的其他建筑遥相呼应，体现人文、信息的沟通和交流，跨向未来。

Located to the north of the Bird's Nest and the Water Cube, the National Convention Center was used as a Fencing Hall and an International Broadcasting Center during the 2008 Summer Olympics. The architectural outline is elegant, and the design of the façade is based on curves found in the eaves of ancient Chinese buildings. The design concept is a modern interpretation of traditional architectural forms. At the same time, it symbolizes a bridge relating to surrounding buildings in the Olympic Park. The building reflects the communication of humanities and information, also represents the future.

北京冬奥村人才公租房

Beijing Winter Olympic Village
Public Housing for Athletes

设计时间 Design Began：2018
建设面积 Building Area：140 000m^2

北京冬奥村位于北京奥体中心南侧，在2022年冬奥会和冬残奥会期间为运动员公寓，赛后转化为北京市高端人才公租房。北京冬奥村规划与建筑设计体现了可持续发展的奥运精神，服务设施尽量由既有建筑改造而成，采用装配式钢结构满足空间的灵活转换，将传统文化与当代生活相结合。同时，智能化无障碍设施在赛时和赛后践行了以人为本的宗旨。

The Beijing Winter Olympic village is located to the south of the Beijing Olympic Center. It will serve as living quarters for athletes during the 2022 Winter Olympic Games and Winter Paralympic Games, and will later be transformed into public rental housing for athletes. The planning and architectural design of the Beijing Winter Olympic Village embodies the Olympic spirit of sustainable development. The service facilities, which include the renovation of existing buildings using fabricated steel structures to meet the flexible space transformation, also combines traditional culture with contemporary design. In the meantime, intelligent barrier-free facilities implement the people-oriented principle.

国家速滑馆

National Speed Skating Oval

设计时间 Design Began：2017

建设面积 Building Area：96 000m²

国家速滑馆是2022年北京冬奥会的标志性工程，也是唯一的新建场馆。建筑以“冰”和“速度”为设计主题，创新的曲面幕墙、超大跨单层索网双曲屋面、集约的全冰面设计，以及智能化的数字运维，呈现出面向未来的“冰丝带”。

The National Speed Skating Oval is the only new venue that will be built, and will be a landmark project for the 2022 Beijing Winter Olympic Games. The design theme derives from the “ice” and “speed”. Innovative curved curtain walls with super large-scale hyperbolic roof span with single-layer cable net, intensive ice ground and intelligent digital operation and maintenance represent the “Ice Ribbon” towards the future.

国家游泳中心改造

Reconstruction of the National Swimming Center

设计时间 Design Began：2018

建设面积 Building Area：140 000m²

合作公司 Partner：Populous

国家游泳中心在2022年冬奥会期间将被改造成冰壶场馆。通过可拆卸场地、室内环境实时控制等技术，将“水立方”转换成为“冰立方”，在冬奥会历史上第一次实现可转换的冰壶比赛场地。同时保留夏奥、冬奥遗产，更好地实现奥运场馆的可持续发展。

The National Swimming Center will be transformed into a curling venue for the 2022 Winter Olympic Games. During this period, the “Water Cube” will be converted into an “Ice Cube” through detachable structure and real-time technological control of the indoor environment. Water Cube will be the first Winter Olympic Stadium to create curling ice in a swimming pool, simultaneously preserving two Olympic venues.

国家速滑馆

National Speed Skating Hall

陆

盛世华章·主场外交建设

HOME COURT DIPLOMACY COMPLEX

建筑经典 讲述中国故事

ARCHITECTURAL CLASSICS TELL THE STORY OF CHINA

十八大以来，中国在全球治理领域积极进取，一个显著的体现就是我们的主场外交活动越来越多，这种主场外交已经成为中国特色大国外交的重要形式。而营建这样重要的交往空间，对于设计者是一个挑战。峰会建筑作为在国际语境下的国家文化的高端呈现，需要在传递友好与开放的同时，平和地表达对中国文化的自信，以及对来宾的尊敬。

2014年，北京APEC峰会在美丽的雁栖湖举办，北京建院设计的以“汉唐飞扬”为主题的雁栖湖国际会议中心，将中国传统文化与现代建筑完美融合，一经亮相便吸引了世界的目光；2016年，杭州G20峰会中，北京建院作为设计总协调再次汇聚全球焦点，以“水墨中国”的思想贯穿始终，尽显大国风范，融合江南特色；2017年，“一带一路”国际合作高峰论坛中，北京建院设计建造的国家会议中心、人民大会堂、国家大剧院、雁栖湖国际会议中心贯穿始终，惊艳全球；2018年，北京建院肩负中国建筑设计的使命担当，再度承接了中国举办的国际级盛会——厦门金砖五国峰会会场的设计服务工作……国家重要发展阶段和举办有国际影响力的大型活动的时刻，都有北京建院的身影；北京建院秉承“建筑服务社会，设计创造价值”的核心理念，服务国家、服务社会、服务大众，创造更大的社会价值与历史价值。

Since the 18th National Congress of the Communist Party of China, more and more diplomatic events have been held at home. Home-court diplomacy has become a main way of major-country diplomacy with Chinese characteristics. It is quite a challenge for architects to design and build such important social spaces. Summit complex, as a sublime demonstration of the best Chinese culture in an international function should convey friendliness and openness, express self-confidence in China's own culture and show respect to all guests.

In 2014, the APEC Economic Leaders Meeting was held in Beijing at the Yanqi Lake International Convention & Exhibition Center. This venue was designed by BIAD with a design theme of “beauty of the Han and Tang dynasties” and it perfectly combined traditional Chinese culture and modern architecture. In 2016, BIAD was the leading design coordinator of the G20 Hangzhou Summit in which it employed the concept of Chinese ink painting to show great manner and Jiangnan-style (south of the Yangtze style). In 2017, during the Belt and Road Forum for International Cooperation, BIAD's China National Convention Center, Great Hall of the People, National Center for the Performing Arts of China, and the Beijing Yanqi Lake International Convention & Exhibition Center amazed the whole world. In 2018, BIAD again was tasked with the mission of designing an international convention center for the Xiamen BRICS National Summit as representative of Chinese architecture design. At every important stage of national development and every moment of international events, BIAD will always work for the country, society and the public with the core belief that “architecture serves society, design creates value”, to create a more social and historical value.

博鳌亚洲论坛会议中心暨索菲特大酒店

Bo'ao Forum for Asia Conference Center

博鳌国宾馆

Bo'ao State Guesthouse

建成时间 Completion Year：2011

建设面积 Building Area: 34 501m^2

博鳌国宾馆地处海南博鳌龙潭岭主峰山麓，近邻博鳌亚洲论坛永久会址，远眺万泉河、龙滚河、九曲江三江入海口，是海南岛独具一格的“山体国宾馆”。

Bo'ao State Guesthouse located in the foothills of the main peak of Longtanling Ridge, Hainan. Neighbouring the permanent venue of the Bo'ao Forum for Asia, Overlooking the estuaries of Wanquan River, Longgun River and Jiuqu River, the guesthouse is a unique "Mountain Guesthouse" on Hainan Island.

博鳌亚洲论坛永久会址二期

Bo'ao Forum for Asia Permanent Venue, Phase Ⅱ

建成时间 Completion Year：2015

建设面积 Building Area：62 840m^2

博鳌亚洲论坛永久会址二期位于博鳌东屿岛会址一期南侧的东屿岛国际会议中心区，项目建成后作为博鳌亚洲论坛年会永久会址的一部分，以一流设施服务高端国际峰会。

Designed as part of the permanent meeting complex of the Bo'ao Annual Forum for Asia, The Phase Ⅱ of the Bo'ao Forum is located to the south of the Dongyu Island International Conference Center of the Phase Ⅰ area, with its first-class facilities and services, has become the preferred location for high-end international summits across the globe.

雁栖湖会议中心：第22届APEC峰会主会场/一带一路国际合作高峰论坛主会场

Yanqi Lake Conference Center (Main Venue of the 22th APEC Summit / Main Venue of the Belt and Road Forum for International Cooperation)

建成时间 Completion Year：2013
建设面积 Building Area：42 000m^2

以“汉唐飞扬双展翼”的主题表达中国在世界舞台上的复兴态势，以建筑形体的北侧契合地形，而在南侧形成围合感的迎宾空间，南北的空间变化以及飞扬舒展的屋檐转角令人印象深刻。

The building image of “Flying Wings of the Han and Tang Dynasties” represents China's return to the world stage. The architectural composition on the north side echoes the terrain, creating an enclosed welcome space on the south. This space variations of the south-north axis and the flying corners of the eaves is very impressive.

杭州国际博览中心改造：第 11 届 G20 峰会主会场

Hangzhou International Expo Center Renovation
(Main Venue of the 11th G20 Summit)

建成时间 Completion Year：2016

建设面积 Building Area：174 700m²

以“廿国共宇同坐轩”为理念，将江南建筑的典雅特质与国家礼仪空间的气势相结合，巧妙利用原有建筑的结构与空间特征，重新组织国际峰会所需的功能流线，重新诠释大尺度园林建筑语言的可能性，表达了时代性与国际化的诉求，展现了中国文化的多样性与亲和力。

The Hangzhou International Expo Center Renovation project combines the elegance of Jiangnan (south of the Yangtze river) traditional architecture with state ritual space. Taking advantage of the structural and spatial features of the original buildings, the project reorganizes different functional circulations required by the function such as summit meetings for heads of States. The refreshed design shows the potential of applying the architecture languages in building a large-scale garden, and appealing China's expression of zeitgeist and internationalization, also the multiplicity and affinity of Chinese culture.

厦门国际会议中心改造：第9届金砖会晤主会场

Renovation of the Xiamen International Conference Center (Main Venue of the 9th BRIC Meeting)

建成时间 Completion Year：2017

建设面积 Building Area：33 000m²

在提炼海洋文化、中国传统文化与福建地域文化元素的基础上，运用现代国际化的语言、柔和含蓄的表达方式，将“国际语言，中国韵味，闽南情怀”的设计主旨，融合在金砖五国峰会会议中心的整体空间中。同时以“丹冠飞羽飘海丝”为创意，提炼闽南文化特征，将“金形山墙”转译成迎宾柱廊，表达开放与包容的态度。

Based on the refinement of maritime culture, Chinese traditional culture and the regional culture of Fujian, along with the employment of an international modern style and a soft and subtle means of expression. The design theme of the renovation project “Chinese charm, and Minnan (southern Fujian) Feelings” was incorporated into the overall space of the BRICS Conference Center. Furthermore, inspired by the theme of the 2017 BRIC meeting “A Red Feathered Phoenix Flies Across the Silk Road of the Sea,” the design utilizes a unique feature of Minnan architecture, the 金 (the Chinese character for gold) shaped gables, re-purposing them to greeting colonnades, creating a space with the atmosphere of openness and inclusiveness.

北京雁栖湖生态发展示范区规划景观综合提升设计

Integrated Improvement Design of Beijing Yanqi Lake Ecological Development Demonstration Area Planning Landscape

柒 走向世界·海外工程建设
OVERSEA CONSTRUCTION

和平发展 走出国门
PEACEFUL DEVELOPMENT AROUND THE WORLD

中国在致力于自身发展的同时，随着国力不断增强，在全球事务中承担了越来越多的国际责任和义务，突出表现在海外建设项目的承接及对外援助项目的建设上。而作为工程设计领域的国企代表，北京建院也正通过不断的项目实践，一步步推进着中国与世界各国之间的平等交流。

从1958年开始承接第一个对外援助成套项目任务开始，北京建院的援外工程史至今已超过60年，已承接了40多个国家的百余项援外工程，其中包括众多的大型民用公共建筑项目：援扎伊尔人民宫、援突尼斯青体中心、援加蓬国民议会大厦、援孟加拉国际会议中心、援科特迪瓦体育场、援塞内加尔国家剧院、援斯里兰卡国家艺术剧院、援老挝人民革命党中央办公楼、援刚果（布）议会大厦、援斯里兰卡国家医院门诊楼、援多哥总统府……大量的援外设计（包括项目管理）任务、可行性研究、设计监理（顾问咨询）任务也为北京建院积累了丰富的援外设计经验。

近些年，在国家积极外交战略的引领下，北京建院人更是积极投身于以“一带一路”为指引的一系列海外建设之中，完成了一批又一批有影响力的建筑作品。在各方的共同努力下，中国梦已打造出开放包容的国际合作平台，促进了全球共同繁荣发展。无论过去、现在和未来，北京建院都始终践行“建筑服务社会，设计创造价值”的核心理念，为祖国的发展和世界的繁荣贡献自己的力量。

While committed to its own development, China's growing national strength is seeing it take up more and more international responsibilities and obligations in global affairs, highlighted by undertaking overseas construction projects and constructing foreign aid projects. As a representative of a state-owned enterprise in the field of design and engineering, BIAD is also trying to promote equal exchanges between China and other countries through continuous project assistance.

Beginning with its first foreign aid project in 1958, BIAD has had a 60-year long history of foreign aid engineering experience in hundreds of projects covering more than 40 countries, including many large civic public building projects. It has provided assistance in the construction of the following projects: the People's Palace in Zaire, the Sports Center for Youth in Tunisia, the National Assembly building in Gabon, the International Convention Center in Bangladesh, the Sports Stadium in Cote d'Ivoire, the National Center for the Performing Arts in Senegal, Sri Lanka National Art Theatre, the Central Office building of the People's Revolutionary Party in Laos, the Parliament House in the Democratic Republic of Congo, the Outpatient Building of the National Hospital in Sri Lanka, and Presidential Palace of the Republic of Togo. BIAD has accumulated a great deal of foreign aid design experience through all these foreign aid projects in which it has been involved in design management, feasibility studies, design supervision and consultation.

In recent years, guided by the national strategy of active diplomacy, BIAD has focused on a series of overseas construction projects that are a part of the Belt and Road initiative and finished a series of architectural projects. Thanks to the efforts of many parties, the Chinese dream is now becoming an open inclusive platform for international cooperation to promote common prosperity and development around the world. BIAD will continue to support the core belief that "architecture serves society, design creates value" to make its own contribution to the development of China and the prosperity of the world.

中国驻英国使馆新馆舍改扩建工程(实施方案)
The Extension of Embassy of the People's Republic of China in United Kingdom(Implementing Plan)

明斯克北京饭店（下图）

Beijing Hotel Minsk

建成时间 Completion Year：2014

建设面积 Building Area：32 924m²

项目位于白俄罗斯首都明斯克市中心的斯维斯洛奇河畔，采用中式院落的围合布局，白墙灰瓦，巧妙地融合了中西建筑元素。同时充分挖掘当地自然文化特征，让建筑以一种平静优雅的姿态融入这座如画般美丽的城市。

This project is located on the banks of the Svisloch River in the city center of Minsk, Belarus. Adopting a Chinese-style traditional courtyard layout with an "enclosed" form, and also with white walls and grey tiles, the hotel skillfully integrates Chinese and Western architectural elements. At the same time, the building takes advantage of local natural and cultural characteristics, and integrates into this picturesque city with a modest and elegant look.

卡塔尔卢塞尔体育场钢结构与屋顶索膜结构工程（右图）

Steel structure and roof cable membrane structure of the Qatar Lusail Stadium

设计时间 Design Began：2017

建设面积 Building Area：183 000m²

合作公司 Partner：Foster + Partners / AFL / Aurecon

卡塔尔卢塞尔体育场是2022年卡塔尔世界杯主体育场，其屋顶是目前世界上最大的屋面膜结构施工项目。北京建院在此项目中进行了复杂的钢结构、索结构和膜结构设计。

The Qatar Lusail Stadium will be the main stadium for the Qatar World Cup in 2022. It will have the largest roofing membrane structure of any stadium in the world. BIAD did the design of the complex steel structure, cable structure and membrane structure for this project.

斯里兰卡国家艺术剧院（下图）
Sri Lanka National Art Theatre

建成时间 Completion Year：2011

建设面积 Building Area：15 562m²

斯里兰卡国家艺术剧院是南亚地区最大的剧院，外观创意取自该国佛教吉祥莲花图案，造型雄伟壮观。通过合理组织屋顶绿化和外墙上的开口，实现遮阳、通风、节能方面的功能。

The Sri Lanka National Art Theatre is the largest theatre in South Asia. The architectural composition is inspired by an auspicious Buddhist lotus pattern. The functions of shading, ventilation and energy savings were achieved a reasonable organization of roof greening and sunshade along with the openings on the facades.

援塔吉克斯坦议会大楼（右图）
Tajikistan Parliament Building

设计时间 Design Began：2018

方案设计 Concept Design：朱古力金 Zukhuritdinov Sirojiddin

建设面积 Building Area：43 000m²

合作公司 Partner：SANOATSOZ

中国援塔吉克斯坦议会大楼和政府办公大楼项目在中国援外史上首次采用对等合作、联合设计模式，是两国平等相待、友好合作的结晶。

One of China's foreign aid projects, the Tajikistan Parliament building adopts the peer-to-peer cooperation and joint design model for the first time in history, which becomes a crystallization of equality and friendly cooperation between the two countries.

多哥共和国总统府

Presidential Palace of the Republic of Togo

建成时间 Completion Year: 2007

建设面积 Building Area：10 000m^2

多哥共和国总统府位于其首都洛美，主体建筑地上两层，地下一层。结合非洲地域文化特点和政府办公建筑的特殊性，总统府体量与立面设计注重虚实对比、材质对比和圆形母题的重复利用。通过H形柱廊和内庭院扩大建筑的尺度感；建筑立面以现代材料表达对非洲地域文化的理解和感受；圆形平面的接见厅巨大的金色穹顶成为整个建筑的视觉中心和精神所在。

The Presidential Palace of the Republic of Togo is located in the capital of Lomé. The main building has two above-ground floors and one underground floor. Combining the regional cultural characteristics of Africa with the characteristics of government office buildings in Togo, the building emphasizes the contrast between void and solid, different materials, and repeating circular motifs. The impression of an enlarged scale was made possible through a H-shaped colonnade and inner courtyard. The façade uses modern materials to express the understanding and feelings of African local culture. The reception hall with a circular plane is the visual center and core spirit of the entire building which is topped a huge golden dome.

缅甸国际会议中心
Myanmar International Convention Center

建成时间 Completion Year: 2010
建设面积 Building Area : 30 000m^2

充分尊重和利用缅甸独特的地域、气候、文化特征，将地域化的生态建筑设计作为缅甸国际会议中心设计的基本点。通过室外绿化庭院与建筑形体的巧妙穿插、符合当地气候条件的室外灰空间形态、结合文化特色的建筑外遮阳系统和细部设计，实现了建筑设计与被动式节能和可持续概念的完美融合。与此同时，正立面深远的挑檐和简洁的柱廊共同营造了庄严典雅的外在气质形象。

Fully respecting and utilizing the unique geographical, climatic and cultural characteristics of Myanmar, the architectural design of this center takes the regional ecological building as a basic concept for the Myanmar International Convention Center. The combination of outdoor green courtyards with architectural volumes, an outdoor grey space in accordance with local climatic conditions, and an exterior shading system and other detailed designs combine to achieve a perfect combination of architectural design and passive sustainable concepts. At the same time, the deep cornices and simple colonnade of the façade create an ideal image of a foreign aid building should have displayed before the eyes of local people.

白俄罗斯标准游泳馆

Republic of Belarus Standard Swimming Pool

捌

和谐共生·城市环境设计
INTERNATIONAL GARDEN EXPO

尊重自然 共铸地球村
RESPECT NATURE TO BUILD A GLOBAL VILLAGE

顺应自然、保护生态的绿色发展昭示着未来。中国人追求人与自然的和谐；合理利用、友好保护以及绿色的发展繁荣也是北京建院追寻的目标。随着时代的发展，我们也越来越关注城市建筑、环境、景观之间的关系，营造良好的生态，实现经济社会可持续发展。

从北京奥林匹克中心区景观，到中国（北京）国际园林博览会；从唐山世界园艺博览会，到延庆世界园艺博览会；从雁栖湖国际会都规划景观提升，到长安街及延长线景观提升、天安门地区环境景观提升……我们在大型博览会项目中树立标杆，向世界传递我们的信念；在城市生活以及环境交汇的每个细微之处，用设计为城市带来更多绿意与便利。我们倡导尊重自然，将绿色价值观打造成为深入人心的人文情怀。

地球是我们赖以生存的家园。我们每一个人都是其中重要的一部分，携手同行，让我们共同建设美丽地球家园，共同构建属于我们的命运共同体。

Green development in accordance with the laws of nature and protecting the ecology is the future. The Chinese people are pursuing harmony between man and nature through rational utilization and protection. Green development and prosperity are also the goals of BIAD. With the passing of time, we are starting to pay more attention to the relationship between architecture, the environment and landscape in our cities to create a better ecology so as to realize the sustainable development of an economic society.

From the central landscape of the Beijing Olympic Park to China (Beijing) International Garden Expo; from Tangshan International Horticultural Exposition to the Yanqing International Horticultural Exhibition; from improvement of planning and landscaping of the Yanqi Lake International Convention & Exhibition Center to the environmental improvement of Chang An Avenue and its extension area, and the environmental improvement of Tian'anmen Square area, BIAD is setting the standard for large-scale expo projects and it's also spreading our philosophy to the world. In every detail where city life meets the environment, we bring more green and convenience to the city through design. We advocate respect for nature and turn green values into a humanistic environment that is deeply rooted in the hearts of the people.

Earth is the home to everyone who lives on it. As an important inhabitant of the earth we should work hand in hand to build our beautiful homeland and a community with a common destiny.

2019年北京世界园艺博览会
Beijing World Horticultural Expo 2019

2019年北京世界园艺博览会国际馆
The International Pavilion of the Beijing World Horticultural Expo 2019

建成时间 Completion Year：2019

建设面积 Building Area：22 000m^2

以“花伞”为元素、“花海”为主题、“长城灰”为色调，秉承“以人为本”“绿色节能”“可持续发展”的设计理念，打造出一座与周边环境和谐共生，既满足展会需求、又满足会后使用的国际馆。国际馆将作为此次世园会三大主场馆之一，承担“以植物和园艺，会八方之友”的主要功能。

Taking the “Flower Umbrella” as a design element, and the “Flower Sea” as a design theme, the pavilion has adopted “Great Wall Gray” as the main color for the structure. In the meanwhile, the concepts of “Energy Saving” and “Sustainable Development” have been integrated into the design, creating an international pavilion that is in harmony with the surrounding environment, and that would meet the needs of the exhibitions and other functional requirements after the conclusion of the Expo. As one of the three main venues of the World Horticultural Expo, the International Pavilion will be the center for gathering people from all over the world through the exhibitions of plants and gardening.

北京植物园展览温室
Beijing Botanical Park Greenhouse

建成时间 Completion Year：1999

建设面积 Building Area：9800m^2

北京植物园展览温室位于植物园中轴路西侧，独巨匠心地以交织的钢结构作为根茎，以起伏的玻璃形态作为绿叶，加上一个椭圆状的花蕾，多种元素的集合，成就了“绿叶对根的回忆”的构想，远远望去宛如西山脚下的一颗明珠。展览温室工程总用地5.5公顷，划分为四个主要展区：热带雨林景区、四季花园景区、沙漠植物景区、专类植物展区。

The Beijing Botanical Park Greenhouse is located on the west side of the central road of the greenhouse. The intertwined steel structure stands for the rhizome, and the continuous waving glass stands for the green leaf, with a flower bud with elliptical shape. All the elements are collected to realize the concept of “memories of green leaves for roots”. From a distance the building looks like a pearl at the foot of Western Mountain. The greenhouse project covers a total area of 5.5 hectares and hosts four main exhibition areas, including a tropical rainforest, four-season gardens, desert plants, and special plants.

中国园林博物馆

Chinese Garden Museum

建成时间 Completion Year：2013

建设面积 Building Area：49 950m^2

中国园林博物馆是中国第一座以园林为主题的国家级博物馆，由主体建筑、室内展园与室外展区三部分组成。博物馆设计以“建筑为纸，园林为画”为理念，并从轴线、院落、天际线、色彩四个方面展示当代中国建筑的“新而中”。利用轴线控制建筑总体布局，在分散的园林化的平面布局中取得整体空间形态。

The Chinese Garden Museum is the first national museum with the theme of traditional gardens, which consists of the main building, an indoor exhibition space and an outdoor exhibition area. The design takes “Architecture as the paper, and gardens as painting” as the concept, with the elements of the axis, courtyard, skyline and colors, to express the characteristic of contemporary Chinese architecture. Taking advantage of a central axis to control the overall layout of building clusters an entire pattern from scattered landscape layout plans has been achieved.

2016年唐山世界园艺博览会

2016 Tangshan World Horticultural Exposition

玖

北京故事·首都城市更新改造
CAPITAL URBAN REGENERATION

传承与发展 复杂问题的新挑战
INHERITING AND DEVELOPING: NEWCHALLENGES TO COMPLEX ISSUES

北京是见证历史沧桑变迁的千年古城，也是展现国家发展新面貌的现代化城市，更是东西方文明相遇和交融的国际化大都市。深入挖掘北京历史文化遗产的内涵和价值，讲好“北京故事”是北京义不容辞的责任。成功的城市更新案例都离不开精心的规划，这些老城保护复兴规划、环境整治提升设计结合历史文化、地域特色、生态环境、产业配置和人居生活等多种要素，为城市的共生发展提供了新动力。

从白塔寺、什刹海到南锣鼓巷，历史保护街区纵横交错的胡同和院落像时空隧道，引人穿越时代的喧嚣，重归曲径通幽、岁月静好；从前门大栅栏北京坊到三里河公园的城市更新设计，用时尚和艺术的魔力重塑城市公共空间，激发老城活力。

怀着对历史的敬畏、对文化的尊重、对城市的热爱，在政府有关部门的支持下，北京建院创新性地采用多部门联动、多专业协同、重点难点专题突破等手段，活跃在北京的“百街千巷”。面对新时代城市规划从增量到存量与减量的转型，我们以全面的技术实力，综合解决复杂的城市问题，全方位提升城市影响力。我们坚持从城市和环境的需求出发，关注建筑与城市空间、经济发展的内在联系和迫切需要解决的问题。一方面，我们以精细化、个性化的设计满足用户需求；另一方面我们也以标准化、装配化的设计适应产业化的发展要求。

Beijing is a historical city that has witnessed thousands of years of change, a modern city showcasing the new face of China's development, as well as an international metropolitan where Eastern and Western civilizations meet and blend. It is tasked on the city of Beijing to deepen the meaning and value of its historical and cultural heritage and to tell a good “Beijing story”. Successful urban regeneration cannot be separated from careful planning. The protection and revitalization plan, and environmental improvement design of these old cities should be recognized for their historical culture, regional characteristics, ecological environment, industrial configuration and residential life so as to drive the symbiotic development of cities.

From Baita Temple, Shichahai, to Nanluogu Xiang, the criss-crossing lanes and courtyards in the protected historical blocks are like time and space tunnels, inviting people to pass through the hustle and bustle of time back to a secluded place in quiet years. From Beijing Fun of Dashilan at Qianmen to the urban regeneration design of Sanli Riverfront Park, the magic of fashion and art rebuilds urban public space and revitalizes old cities.

Holding a reverence for history, respect for culture and love for our cities, and with the support of relevant government departments, BIAD has innovatively adopted inter-departmental collaboration, inter-disciplinary cooperation, and key and difficult topics breakthrough, among other methods, to actively participate in projects related to the streets and lanes of Beijing. Faced with the transition of urban planning from increment to stock and to reduction in the new era, we have developed our technical strength to comprehensively solve complex urban problems and enhance the overall urban influence. We persisted on starting from the needs of the city and environment and focused on the internal relations between architecture and the economic development of urban spaces and urgent problems. We have employed a fine and personalized design to meet user demands. On the other hand, we have adapted to the development of industrialization with standardized and assembled design.

北京坊城市更新

Urban Regeneration of Beijing Fun

全国妇联办公楼改扩建工程
Renovation and Expansion of the National Women's Federation Office

建成时间 Completion Year：2015

建设面积 Building Area：35 772m²

在保证不拆除原有主体办公楼的前提下进行改建。北侧、南侧分别扩建为北办公楼和礼仪大厅。在尊重妇联历史和所在地段建筑风格的基础上进行优化设计，与周围环境协调统一，体现沉稳、端庄、包容的设计理念。扩建后建筑群体近似“工”字形，在东侧围合出一个院子，为日常人流、车流提供回转空间。功能布局合理，内外空间尺度适宜，整体效果、细节控制较好，品质明显提升。

This renovation and expansion project were carried out based on the condition that the original structure be maintained and that the main office building would not be demolished. The north and south areas have been expanded into an office building and a ceremonial hall respectively. The optimized design is based on respecting the history of the Women's Federation and regional architectural styles, which is a response to the surroundings, and which also reflects the design concept of modest, dignified and inclusive features. After the expansion, the building cluster was similar to the shape of the Chinese character, which is enclosed in the east side as a courtyard, providing sufficient space for the circulation of people and transportation. The functional layout is reasonable, and both the internal and external spatial scales are friendly, with the overall effect outstanding with perfect details design.

华电（北京）热电有限公司天宁寺厂区
Huadian (Beijing) Thermal Power Plant in Tianning Temple

建成时间 Completion Year：2016

建设面积 Construction Area：20 707m^2

改造设计以延续场所记忆，协调厂区与周边关系为出发点，力图修旧如旧，渐进式发展，重整厂区内部空间，梳理多种尺度各异的建筑，并重塑厂区与天宁寺及现代城市的关联，为远期建设提供基础。

The renovation design of the Thermal Power Plant in the area of Tianning Temple attempts to continue to retain the venue style and coordinate the relationship between the plant complex and its surroundings. We have striven to retain the heritage as much as possible through progressive development, to renovate the inner space of the plant complex and buildings of different scales. The project has also reorganized the connection between the plant area and Tianning Temple and a modern urban space, providing a basis for long-term construction.

北京什刹海鼓楼西大街复兴工程
The Renovation and Revival Plan for West Gulou Street in Beijing

设计时间 Design Began：2017
规划面积 Planning Area：$1km^2$

基于鼓楼西大街街区特点，整理与复兴计划以高品质风貌休闲区为规划定位，颠覆传统改进式治理思路，立足于街区现存问题，在解决问题的基础上形成该地区新的发展动力，探索在城市治理方面新的合作模式。鼓励居民、企业等建立与地方政府的合作与伙伴关系，有效动员、引导社会参与，推动共治共享，激发社会责任意识，解决地区现存问题，实现"乱象治理+品质提升"。

Based on the characteristics of Guxi Street, this plan aimed to take this high-quality leisure space as its orientation, and transform the traditional means for governing, and based on the existing problems of the street, the new development motivation is created to solve all the problems and explore a new cooperation mode in urban governance. We have encouraged residents and enterprises to cooperate with the local government to mobilize and guide social participation, also promoting co-managing and sharing, to stimulate the sense of social responsibility. In this way, existing problems will be solved, and the chaos will be controlled, and the quality of the living environment will be improved.

王府井街道整治之口袋公园
Renovaiton of the Pocket Park of Wangfujing Street

建成时间 Completion Year：2018

建设面积 Construction Area：950m²

通过对空间的划分等将城市空间的复杂性呈现出来。“墙上痕”通过造型和材料，暗示了新与旧的对话。树木以及“墙上痕”上的立体绿化，共同构成了一幅幅老北京四合院内旧时生活场景的片段。

By the means of space division, the complexity of urban space is presented. The design of the park suggests the dialogue between new and old through shapes and materials. The trees and the three-dimensional greening on the man-made marks on the wall create a series of living scenes of Beijing traditional quadrangle dwellings during olden times.

北京西打磨厂 222 号院
Beijing West Grinding Factory NO.222

建成时间 Completion Year：2016

建设面积 Construction Area：337m²

拆除东跨院当中的小房子，在原来墙体的位置代之以清水混凝土墙体，标识出原有建筑的记忆。在其上部南北两侧挑出钢梁，跨越在南北两个房子的顶部之上，形成一个桥形装置。

Demolishing the small building in the east yard, and replacing the original wall with pure concrete, to identify the memory of the original building. Steel beams are suspended from both sides of the north and the south, spanned over the top of the two houses, to form a bridge-shaped volume.

前门东区城市更新

Qianmen East Urban Regeneration

拾

面向未来·设计的创新与突破
DESIGN INNOVATION AND BREAKTHROUGH

展现中国智慧 积极拓展
UTILIZE CHINESE WISDOM AND EXPAND ACTIVELY

当今世界，新一轮科技革命蓄势待发，人工智能时代已经来临，一些重大科学问题的原创性突破正在开辟新前沿、新方向，一些重大颠覆性技术创新正在创造新产业、新业态，产业更新加快，社会生产和消费从工业化向自动化、智能化转变，生产率将产生重大飞跃。对比今天和十年前，建筑界又有了全新的发展，这种发展是跨越式的。

在建筑业中，“中国制造”正逐步走向“中国创造”。作为其中重要一分子的北京建院也正积极利用数字技术指导全过程设计控制与优化，形成了一系列科技创新技术：首创性地构建了适用于高精度控制要求的数字技术平台，利用创建高质量建筑信息模型与数据库，实现了对所有建筑可视面的优化；通过创造性地构建建筑几何控制系统，精确解决了当代复杂建筑设计控制中建筑系统之间的定位“拼合”问题。

通过大数据、云平台、新的算法、性能化的设计，北京建院连通了由设计到施工的全程。而经过多年“开行业之先”的建筑工程积累，北京建院也开启了结构设计在天文科技领域的突破，成功跨入了大科学仪器设计领域。500m口径射电天文望远镜之后，我们又继续助力打造江门中微子实验中心探测器结构工程，为国家和民族的基础科研事业提供坚实保障。

探索和创新始终是北京建院的文化基因。从外国人原创、中国人配合，到完全由中国建筑师原创与深化设计、由中国技术实现设计，北京建院正在用一系列实践证明自己的实力。

In today's world, a new round of technological revolution which poised for growth as the era of artificial intelligence is upon us. Breakthrough in solving some important scientific problems is opening up new frontiers and new directions. Some major innovations in revolutionary technologies are creating new industries and new programs. The upgrading of industry is happening faster, with social production and consumption changing from industrialized to automatic and intelligent. Productivity is poised to have a significant leap forward. Compared with 10 years ago, the architecture industry today is experiencing a new, leap-frog development.

In the architecture industry, “Made in China” has evolved into “Create in China” step by step. We are playing an important role, actively using digital technology throughout the design process of control and optimization, and we have formed a series of scientific and technological innovation technologies. First, we constructed a digital technology platform suitable for high-precision control requirements. Then we realized the optimization of all visual aspects of buildings by creating high-quality building information models and databases. And then we precisely solved the problem of positioned “integration” among architectural systems in the design control of contemporary complicated architecture.

Through big data, cloud platforms, new algorithms and performance-based design, BIAD has connected the entire process from design to construction. After years of accumulation of architecture project experience “ahead of the industry”, we also made a breakthrough in structural design in the field of astronomical science and technology, successfully entered the large scientific instruments design field. After successfully built the 500m caliber spherical radio telescope, we moved to build the detector structure engineering of Jiangmen Neutrino Experiment Center, providing a solid base for the country and national basic scientific research.

Exploration and innovation have always been the cultural genes of BIAD. We have progressed from a system in which architecture works were originally initiated by foreigners with just the cooperation of Chinese, to completely originated and designed by Chinese architects. With the realization of Chinese technology, BIAD is proving its strength with a series of architecture design works.

丽泽 SOHO
Lize SOHO

凤凰中心
Phoenix Center

建成时间 Completion Year：2013
建设面积 Building Area: 62 800m^2

凤凰中心项目的造型取意于“莫比乌斯环”，由于独特的创意构思，设计团队利用数字技术指导全过程设计控制与优化：构建高精度控制要求的数字技术平台，通过建筑信息模型与数据库，实现了对建筑可视面的优化；构建建筑几何控制系统，解决了建筑系统之间的定位“拼合”问题；结构设计采用“双向叠合网格结构体系”，首创“弥合自由曲面的单向非连续折板幕墙体系”，用平板幕墙弥合大曲率自由曲面，并在加工生产全过程中实现了与数字化设计控制的无缝对接。

The architectural composition of the Phoenix Center is based on the “Mobius Ring”. In accordance with this unique creative idea, our designers apply digital technology to guide the control and optimization of the whole design process to build a digital technology platform with high precision control requirements and realize the optimization of building visual surfaces through information models and databases; construct a building geometric control system to solve the problem of positioning “fitting” between different systems; use “two-way superimposed grid system” for structural design and pioneer the“oneway and non-continuous folded curtain wall system for bridging free-form surfaces”, and use flat plates curtain wall to bridge the large-curvature free-form surface, as well as realize seamless connection with digital design control for the whole process.

500m口径球面射电望远镜结构工程（右图）

The 500-meter Aperture Spherical Radio Telescope (FAST)

建成时间 Completion Year：2015

FAST 项目是“十一五”国家重大科技基础设施，是由随地势布置的格构柱和500m内径圈梁构成的巨型索网系统工程。成型精度需达到毫米级，可实现主动变位、实时调整形态，在观测方向形成300m口径瞬时抛物面以汇聚电磁波。北京建院在设计、加工制作、施工建造等环节完成多项科技创新成果。

The FAST was a major national project of science and technology infrastructure during the 11th Five-Year Plan (2007—2012). It is a huge cable net system composed of columns arranged along with the terrain and 500-meter inner diameter ring beams. The moulding accuracy needs to reach millimeter, which can realize active displacement, real-time shape adjustment, and form a 300-meter calibre instantaneous paraboloid in the observation direction to gather electromagnetic waves. BIAD has achieved scientific and technological innovations for design and construction processing.

江门中微子实验中心探测器主体结构（下图）

The Jiangmen Underground Neutrino Observatory (JUNO) Center,
Stainless-steel Main Structure of Detector

设计时间 Design Began：2016

江门中微子实验是由中国主持的大亚湾中微子实验的二期工程，中心探测器主体不锈钢结构属于国家重大科技基础设施建设项目。中心探测器主体位于地下700m的洞室中，由直径35.4m的充满闪烁体的有机玻璃球和直径40.1m的不锈钢网壳组成，其规模为世界之最。

The Jiangmen Underground Neutrino Observatory (JUNO) is a multipurpose neutrino experimental phase II project hosted by China. The central detector stainless-steel structure is a major national project of science and technology infrastructure construction. The experiment detector is located in a 700-meter underground space, which is composed of a plexiglass sphere with a diameter of 35.4m and a stainless-steel mesh shell with a diameter of 40.1m, which is the largest such structure in the world.

凤凰中心

Phoenix Center

公共建筑
PUBLIC BUILDING

文化 / 体育建筑
CULTURE / SPORTS BUILDING

96 深圳文化中心 Shenzhen Cultural Center
100 深圳湾体育中心 Shenzhen Bay Sports Center
104 故宫北院 North Annex of the Palace Museum
106 首都图书馆新馆 New Extension of the Capital Library (Beijing)
108 炎黄艺术馆 Yanhuang Art Museum
110 国家美术馆 The National Gallery of China
112 嘉德艺术中心 Jiade Art Center

观演 / 博览建筑
PERFORMANCE / EXPO BUILDING

114 中国国际展览中心 2 ~ 5 号馆
Hall 2–5 of the China International Exhibition Center
118 又见五台山剧院 Encore Mount Wutai
122 北京动物园大熊猫馆 The Panda House of the Beijing Zoo
124 珠海大剧院 Zhuhai Opera House
126 中国科学技术馆 China Science and Technology Museum
128 中国电影博物馆 China National Film Museum
130 中国国际展览中心 I 期
China International Exhibition Center Phase I
132 国家大剧院 The National Center for the Performing Arts of China
134 第九届中国（北京）国际园林博览会主展馆 Main Exhibition Hall of the 9th China (Beijing) International Garden Expo

教育 / 科研建筑
EDUCATION / RESEARCH BUILDING

136 中国美术学院南山校区 Nanshan Campus of China Academy of Art
140 北京建筑大学 6 号综合服务楼
Service Building No. 6 of the Beijing University of Architecture
144 北川中学 Beichuan Middle School
148 望京科技园二期 Wangjing Science and Technology Park Phase II
150 联想园区 C 座 Block C of Lenovo Park
152 联想研发基地 Lenovo Research and Development Base
154 腾讯北京总部 Tencent Beijing Headquarters
156 小米科技北京总部 Xiaomi Technology Headquarters Beijing
158 低碳能源研究所及神华技术创新基地 Low Carbon Energy Research Institute and Shenhua Technology Innovation Base
160 国电新能源技术研究院
Guodian New Energy Technology Research Institute

办公 / 商业建筑
OFFICE / COMMERCIAL BUILDING

162 中国石油大厦 China Petroleum Building
166 侨福芳草地 Parkview Green
170 北京城市副中心行政办公区
Administrative Offices of the Beijing City Sub-Center
174 北京奥体商务南区 OS-10B 城奥大厦
South Olympic Business District OS-10B Cheng'ao Building
178 全国人大机关办公楼
Administrative Office of the National People's Congress(NPC)
180 商务部办公楼改造 Ministry of Commerce Office Building Renovation
182 北京国际金融大厦 Beijing International Finance Building
184 北京市高级人民法院 Beijing Higher People's Court
186 恒基中心 Henderson Center
188 北京电视中心 Beijing TV Center
190 王府中环 Wangfu Central
192 蓝色港湾 Blue Harbor
194 法国驻华大使馆新馆 New Embassy of France in China

酒店 / 会议建筑
HOTEL / CONFERENCE BUILDING

196 三亚太阳湾柏悦酒店 Park Hyatt Sanya Sunny Bay Resort
198 昆仑饭店 The Kunlun Hotel
200 钓鱼台国宾馆会议中心
Conference Center of Diaoyutai State Guesthouse

医疗 / 交通建筑
MEDICAL / TRANSPORTATION BUILDING

202 昆明长水国际机场航站楼
Kunming Changshui International Airport Terminal Building
206 广州南站 Guangzhou South Railway Station
210 首都医科大学附属北京天坛医院
Beijing Tiantan Hospital, Capital Medical University
214 深圳宝安国际机场 T3 航站楼
Shenzhen Baoan International Airport Terminal 3
216 青岛北站 Qingdao North Railway Station
218 南京南站主站房 Main Station of the Nanjing South Railway Station
220 北京儿童医院门诊楼 Beijing Children's Hospital Outpatient Building
222 首都医科大学宣武医院改扩建一期工程
Xuanwu Hospital Capital Medical University
Reconstruction and Extension Phase I

深圳文化中心
Shenzhen Cultural Center

建成时间 Completion Year：2007
建设面积 Building Area：89 744.8m²
合作公司 Partner：Arata Isozaki & associates

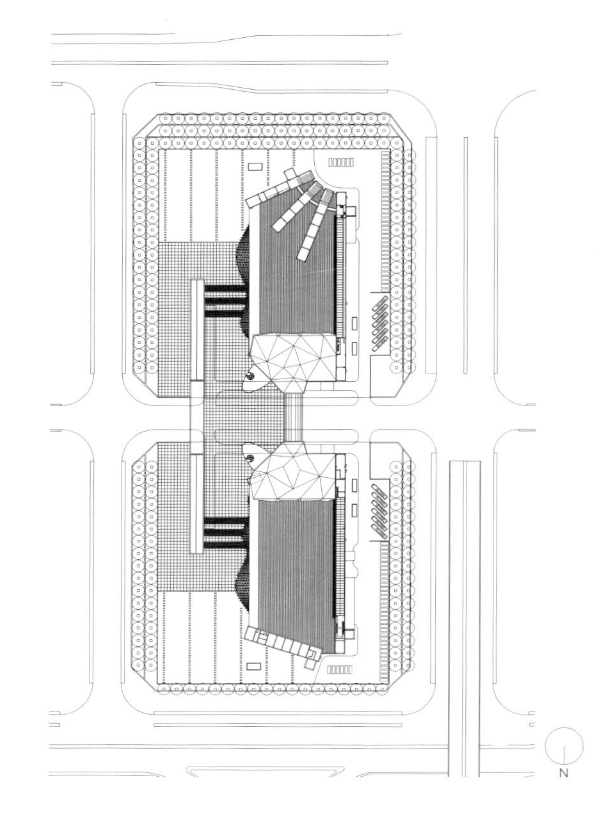

深圳文化中心由中心图书馆和深圳音乐厅两部分组成，两栋建筑之间是将二者连成一体的公共文化广场，形成一个面向市政厅—中央绿化带—莲花山的开放空间。音乐厅和图书馆入口处的两个钢结构玻璃体分别被称为“金树”和“银树”，寓意“文化森林”。中心图书馆的设计藏书量为 500 万册。阅览室为无墙开放空间，共 6 层。各层实现自然采光和通风，并铺设地面线槽以适应书架、桌椅等的变化更新。报告厅分为两层，设置了升降台、投影机、同声翻译等设备。报告厅和休息厅之间通过敞开前室门可形成一个用于聚会等的多用途空间。

深圳音乐厅主要由演奏大厅和多功能小厅组成，大厅由不规则折叠墙面围成，小厅设在入口大厅上方，是以轻音乐为主的多功能音乐厅。入口大厅周围设有向市民开放的咖啡厅、音乐书店等设施。演奏大厅采用峡谷梯田式，厅内采用了座椅送风系统，管风琴的设计采用外露手法。多功能小厅设计成多层平台和升降舞台，既可作为舞台又可作为观众席来使用。

深圳市儿童医院

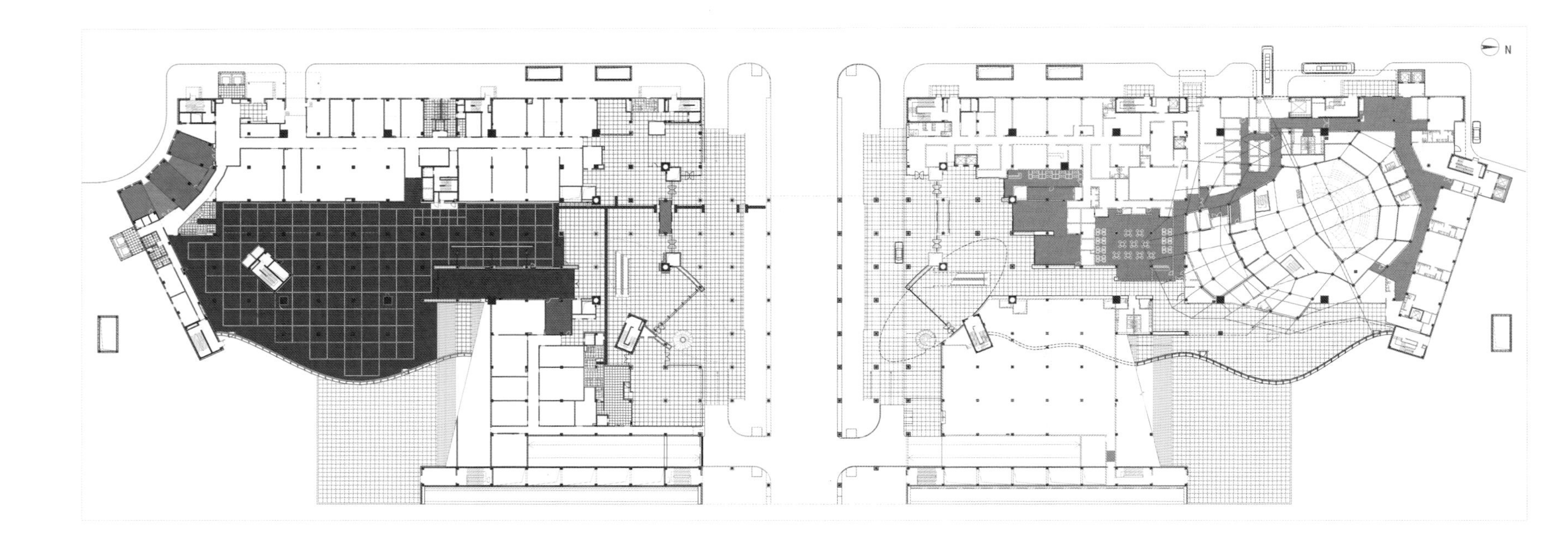

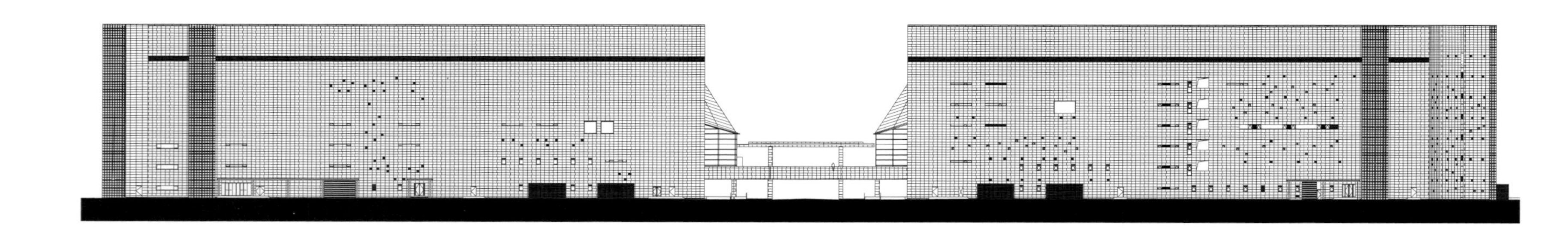

The Shenzhen Cultural Center consists of two parts, the Central Library and the Shenzhen Concert Hall. The two buildings are connected by a public cultural square, which forms an open space facing the city hall, the central green belt and Lotus Mountain. The two steel-structured glass volumes at the entrance to the concert hall and library are called "Golden Tree" and "Silver Tree" respectively, representing the idea of "cultural forests". The central library has a collection of 5 million books and a 6-story open reading space without walls. Each floor is naturally lighted and ventilated, laid with ground troughs to accommodate the redeployment of bookshelves, tables and chairs. The lecture hall is divided into two floors, equipped with lifting platforms, projectors and rooms for simultaneous interpretation. The front lobby between the lecture hall and the lounge can be opened up for parties and other purposes.

The Shenzhen Concert Hall is mainly composed of a performance hall and a smaller multi-functional space. The performance hall is enclosed by irregular folding walls, the smaller space perched above the entrance hall, functioning as a multi-purpose concert hall for light music. Surrounding the entrance hall are cafes and music and bookstores that are open to the public. The performance hall is designed like a terraced field, with a ventilation system installed below the seats; the pipe organ is completely exposed. The multi-functional space is designed with a multi-level platform and a lifting stage, which can be used as a stage or as an auditorium.

深圳湾体育中心
Shenzhen Bay Sports Center

建成时间 Completion Year：2010
建设面积 Building Area：256 520m^2
合作公司 Partner：株式会社佐藤综合计画 AXS SATOW ING

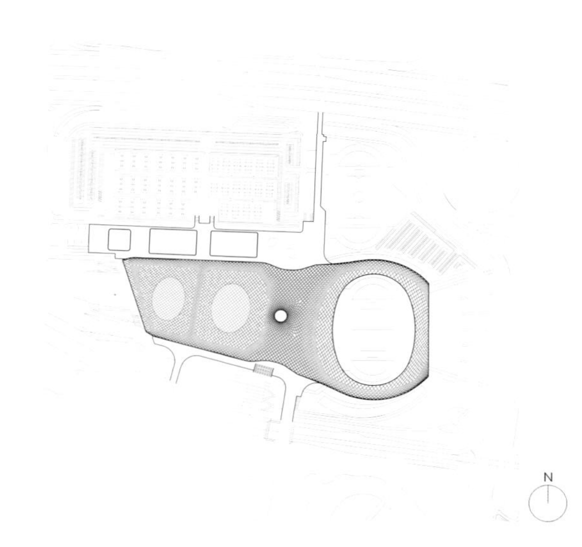

作为 2011 年世界大学生运动会开幕式主场馆，该项目创意为城市“春茧”。530m 超大跨度的银白色单层网壳将体育场、体育馆和游泳馆覆盖其下，集约式的布局节省出大量的室外运动场地。线条柔美的网格屋盖犹如孕育破茧而出、冲向世界的运动健儿的孵化器。

在临海一侧，设计师大胆地“切”出了一个巨大的开口，犹如一个开放的舞台，将自然海景引入场内。开口上部有一座横跨 120m 的天桥，可以在此喝着咖啡观赛赏海，坐在体育场里看大海！国内首创的“海之舞台”，成为建筑与自然相融合的神来之笔。

春茧的造型形成了良好的场馆通风效果。数以万计的天沟收集到的雨水被用于绿化灌溉。海水源热泵、太阳能等组成了合理高效的绿色能源系统。

As the main venue for the opening ceremony of the 2011 Universiade, this project was designed under the concept "Spring Cocoon", in which the stadium, the gymnasium and the swimming pool were all covered by a silver-white single-layer reticulated shell spanning 530 meters. The compact layout saves a lot of outdoor space for sports. The reticulated roof with gentle curves looks like an incubator where the nation's athletic talents are bred and trained.

The architect "cut" a huge opening at one end where the structure faces the ocean, creating an open stage that introduces the natural waterscape into the building. A 120-meter bridge spans the opening, where people can sit in the stadium and enjoy both a cup of coffee and the sea view. This is the first "stage over the sea" in China, which has become an amazing place where architecture and nature are merged together.

The shape of the spring cocoon helps to optimize the ventilation of the stadium, and rainwater collected by tens of thousands of gutters is used for green irrigation, together with the seawater-sourced heat pump and solar energy, in which an efficient and functional green energy system is formed.

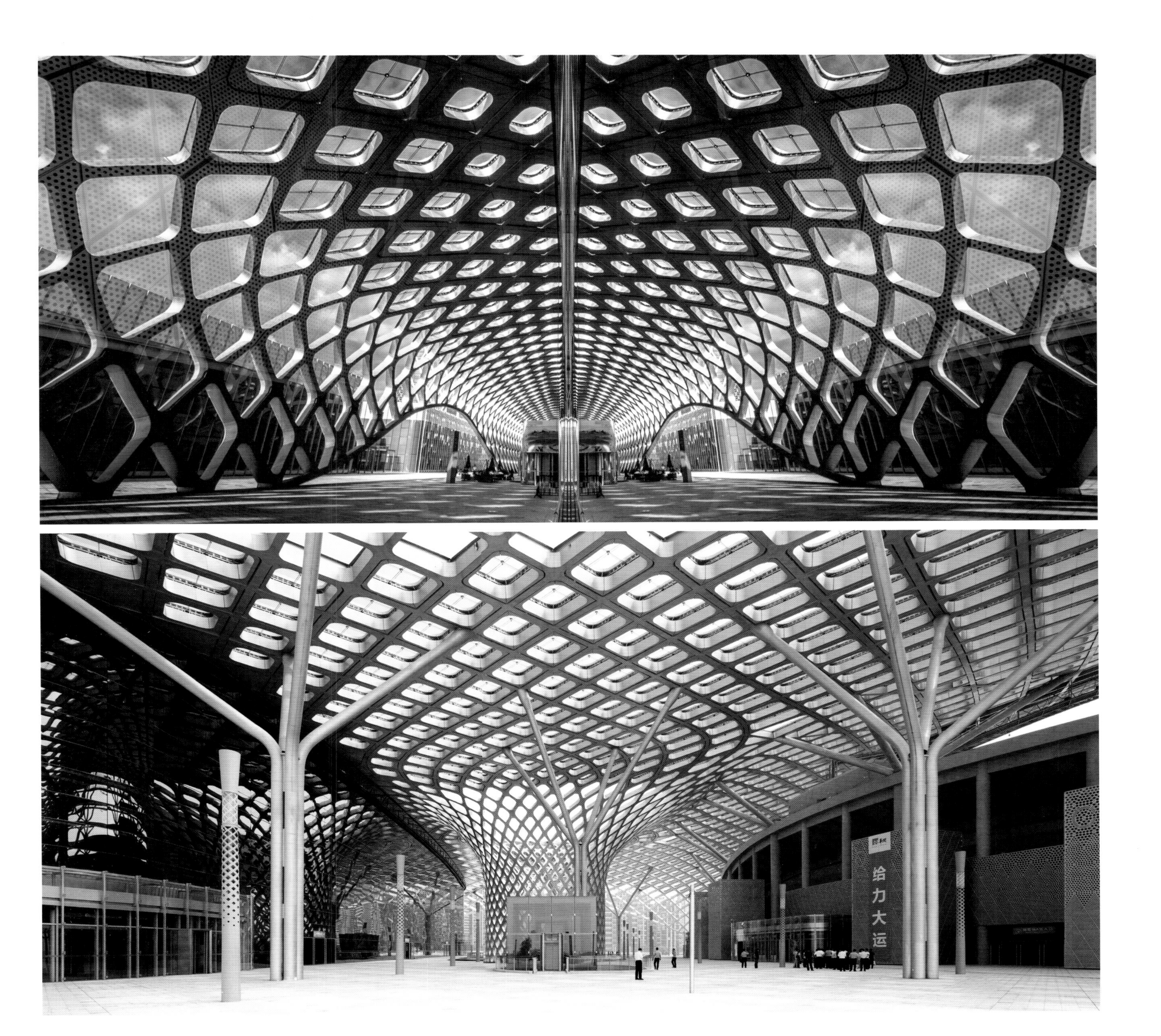
给力大运

故宫北院
North Annex of the Palace Museum

设计时间 Design Began：2015
建设面积 Building Area：102 000m²

建筑布局方面，将对外展览功能布置在地块东侧，较为私密的办公修复及后勤功能布置在西侧，最大限度满足功能布局的同时，也有利于分期建设，同时方便与现有修复用房重新整合。设计灵感来自对中国传统“殿”“堂”“舍”“院”的理解，借助从中提炼的轴线、秩序、等级等特质，形成博物馆群落。从故宫中轴线的高潮——太和门至三大殿中提取尺度关系，完好顺应基地边界。作为整个建筑体量的中枢，主要展览空间——基本陈列展厅被置于中轴线上。

The space for external exhibition is allocated to the east of the site, while private offices and back of house areas are in the west. Such planning helps to maximize the functional space and to develop the project in phases, while reorganizing current spaces for restoration. The design was inspired by the traditional Chinese "palace", "hall", "house" and "yard", from which we extracted the concept of an axis, order and grade to define different clusters within the museum. We explored the relations of different scales from Taihemen to the three major palaces, which cover the most important area in the central axis of the Forbidden City, to define the boundaries of the site. Major exhibitions were basically arranged along the central axis.

首都图书馆新馆

New Extension of the Capital Library (Beijing)

建成时间 Completion Year：1999
建设面积 Building Area：37 000m^2

首都图书馆是综合性大型公共图书馆，位于朝阳区南磨房华威桥东南侧，地上 8 层，地下 1 层。设置各类阅览室 17 个，阅览座位 1000 余个，设计藏书量 400万册。建筑平面取 1/4 扇形，围合中心广场，形成强烈的聚合力。正立面恰似一展开的长卷，带来强烈的书卷气息。主立面中央为一巨大的牌楼造型，成为整个建筑的核心和统率。主入口处玻璃幕墙形成的中国建筑殿堂形象正是国子监大殿辟雍的剪影。辟雍的形象既代表首都，又代表知识的大门。

The Capital Library is a comprehensive large-scale public library located on the southeast side of the Huawei Bridge in Nanmofang, Chaoyang District, with a basement and eight floors above the ground. The library has 17 reading rooms for different categories of books, more than 1,000 reading seats and a collection of 4 million books. Taking the shape of a quarter of a fan, the building encloses a square in the center, which has become an attraction to the communities around. The front elevation looks like an unfolding scroll of a book, creating a strong academic atmosphere. A huge archway is placed in the center of the main elevation to define the core of the entire building. The image of the Chinese architectural hall formed by the glass curtain wall at the main entrance is the silhouette of the ancient imperial university, which represents the identity of the capital and the entrance to knowledge.

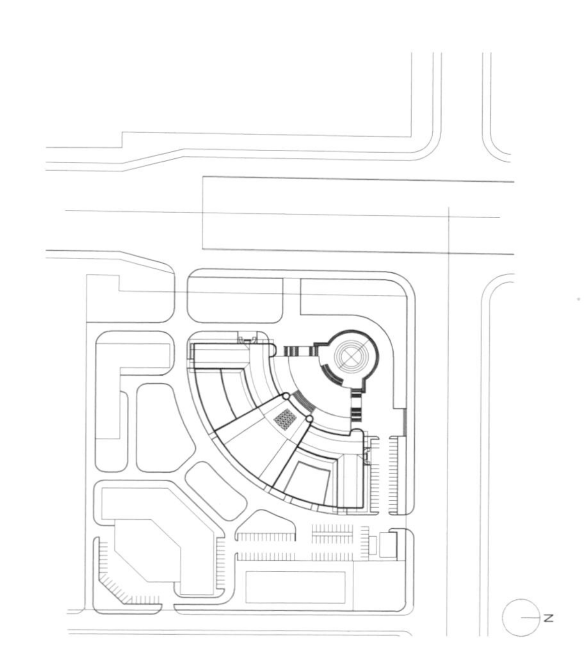

吉林圖書館

炎黄艺术馆
Yanhuang Art Museum

建成时间 Completion Year：1992
建设面积 Building Area：13 240m^2

炎黄艺术馆以簇集的形式组织展览空间，根据展出需要，可分可合，并有部分展厅可直接对外服务。由于体型上小下大，可积极利用顶光，厅内最大限度利用自然光线的同时避免直射光照损害展品。建筑形象运用“斗形”造型。外立面在“斗形”的斜墙面上贴深茶色上釉瓦。半地下室墙面用自然花岗石贴面，以体现基座的深沉、浑厚之感。直墙部分采用青石板。学术报告厅和展室以门厅为中心东西向排列；展室相对集中于二层和三层。

The galleries in Yanhuang Art Museum are organized in clusters, which can be connected or separated according to the different needs of exhibitions. Some galleries are open to the public directly. The ladder-shaped section maximized the natural lighting while protecting the exhibits from direct sunlight. The building, which is in the shape of a ladder, is covered with a dark brown glazed tile on its slanted exterior walls. The walls in the semi-basement are made with natural granite to express the idea of a solidness and thickness of a base structure. The straight walls adopt bluestone planes. The academic lecture hall and the exhibition hall are arranged in an east-west direction from the entrance hall, while the exhibition rooms are mostly placed on the second and third floors.

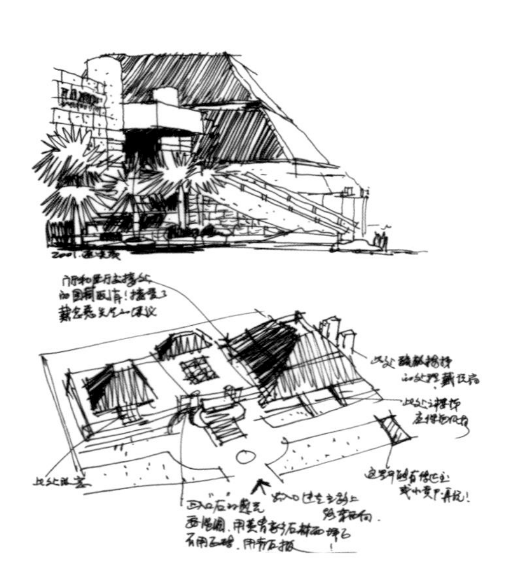

国家美术馆
The National Gallery of China

设计时间 Design Began：2009
建设面积 Building Area：128 600m²
方案设计 Concept Design：Ateliers Jean Nouvel

国家美术馆的建筑形态构思出于毛笔的一划。“一” 是周易卦象之始、中国字的基本元素、中国书法笔划书写的基础。外立面采用不同质感的石材、不锈钢板和丝网印刷玻璃拼合，三种材料相互渗透、衔接，浮现出气韵生动、虚实相间、明暗变化的中国山水画意象。国家美术馆首层架空，地上 6 层，地下 2 层。主要建筑功能包括供人流集散的首层夏季广场，为展览服务的二层冬季大厅、三至六层和屋顶层，以及相关配套和公共服务设施。

This architectural form, which originated from the National Gallery of China, is inspired by the Chinese character “一”(one), which is a basic element in the writing of Chinese characters. The facade of the building is a collage of different materials of different textures, including stone, stainless steel and screen-printed glass. The transition and penetration of the three materials echo the concept of void and solid, of weak and strong, and light and shadow in Chinese Shan Shui paintings. The first floor of the National Gallery is an open space with high ceilings, with 6 floors above ground and 2 below ground. The second floor, also known as the Winter Hall, is for exhibitions. Floors 3—6 and the roof floor were reserved for hockey and exhibitions, facilities for auxiliary and public services were also located on those floors.

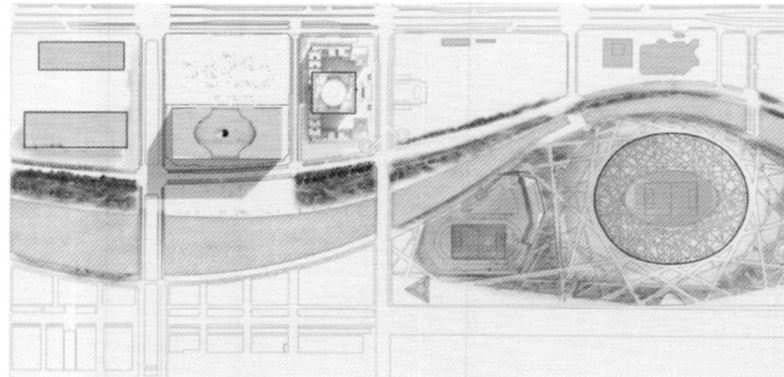

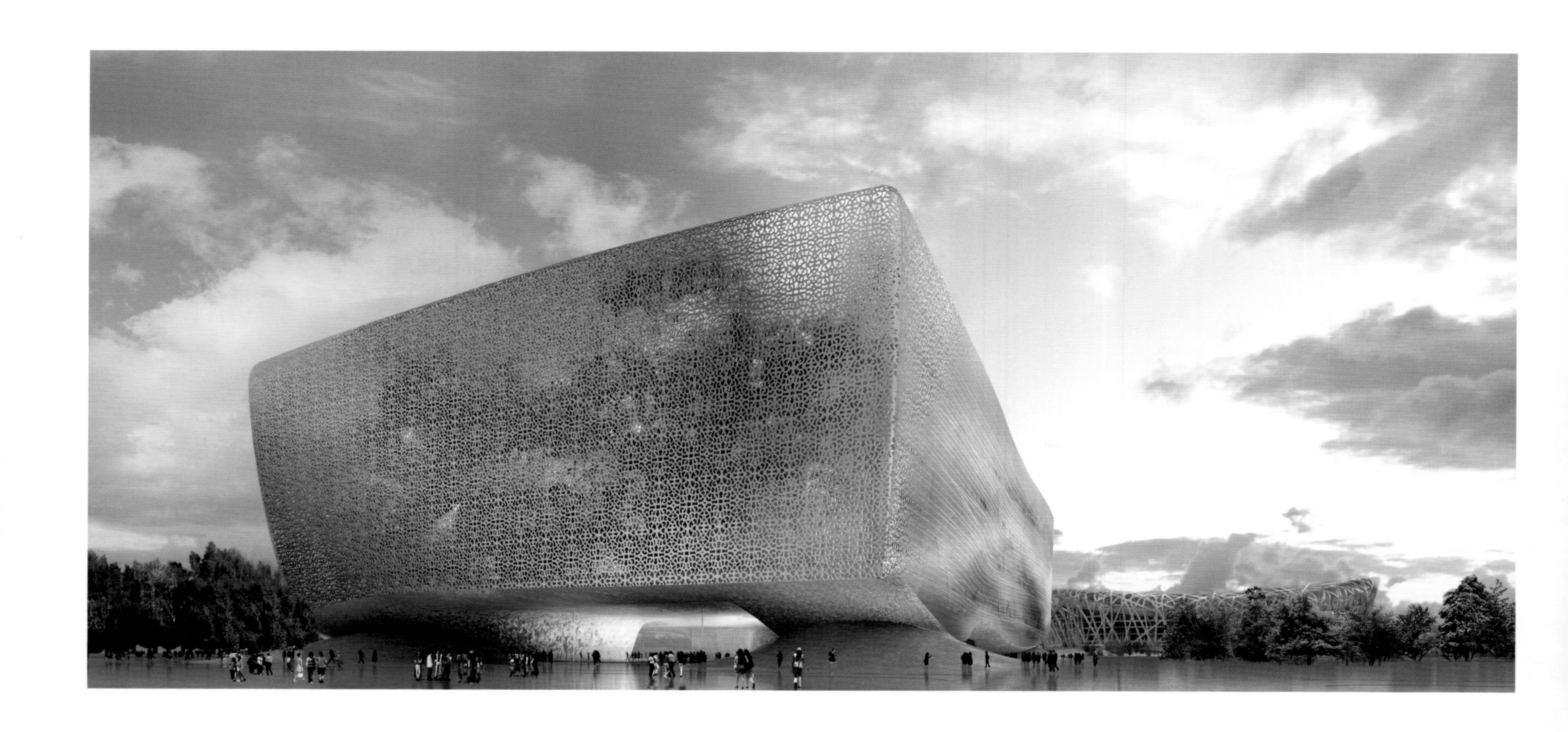

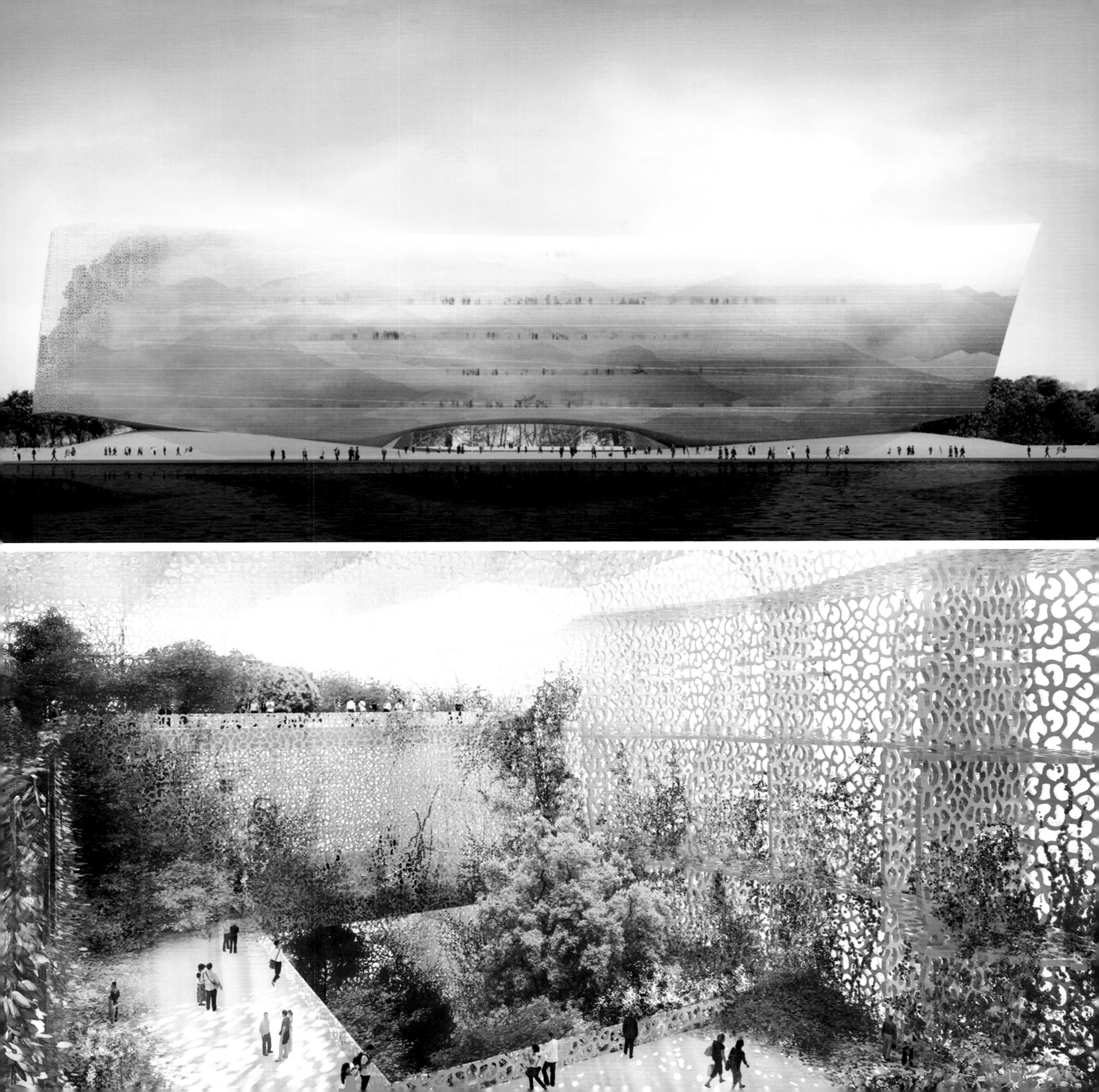

嘉德艺术中心
Jiade Art Center

建成时间 Completion Year：2017
建设面积 Building Area：55 987m²
合作公司 Partner：Buro Ole Scheeren

嘉德艺术中心位于北京市东城区王府井大街1号，是一个集艺术品拍卖、展览、文物储藏、鉴定修复、学术研究、信息发布、精品酒店为一体的亚洲首个“一站式”文物艺术品交流平台。建筑上部四层至八层悬浮的“方环”呼应周边现代建筑，承载酒店功能，无缝服务下部的拍卖展示空间，并采用“透明砖墙”，融入传统建筑中的砖墙肌理。建筑下部将胡同的肌理延伸至基地，采用退台的层叠方式呼应西侧旧城保护区。立面将《富春山居图》抽象到石材幕墙上。

Located at No.1 Wangfujing Street, Dongcheng District, Beijing, Jiade Art Center is the first “one-station” cultural relics and artwork exchange platform in Asia. It integrates artworks auction, exhibitions, cultural relics storage, appraisal and restoration, academic research, information release and a boutique hotel. The suspended “square ring” on the upper four floors of the building responds to the surrounding modern buildings, carrying the functions of the hotel, serving the auction and display space of the lower space successively. The “transparent brick wall” has been adapted to the suspended ring, which is integrated into the brick wall texture of traditional buildings. The lower space of the building extends the atmosphere of a hutong (Chinese alleyway) to the ground floor. At the same time, the stepping back platforms echo the ancient reserve area on the west. On the facade, a famous Chinese painting *Dwelling in the Fuchun Mountains* is abstracted onto the stone wall.

中国国际展览中心 2 ~ 5 号馆
Hall 2–5 of the China International Exhibition Center

建成时间 Completion Year：1985
建设面积 Building Area：25 000m^2

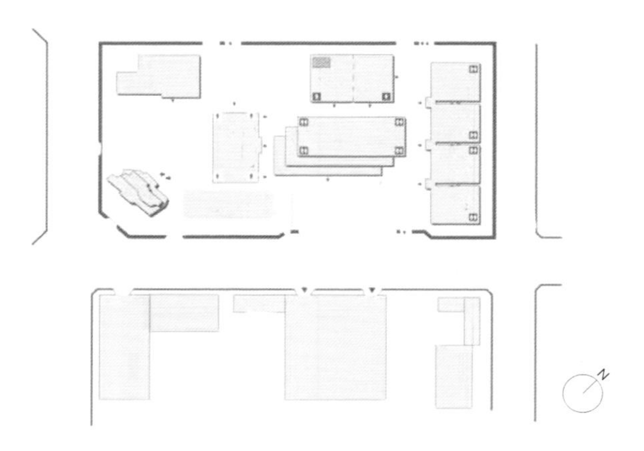

中国国际展览中心 2~5 号馆为一期工程。为适应展览空间灵活多变的功能要求，每个展馆大厅由 63m×63m 的正方形大空间组成；展厅内在标高 4.5m 处沿周边网架柱挑出展廊平台，不仅扩大了展出面积，也丰富了室内外空间。

4 个展厅之间由 3 个连接体相连。作为展览馆的主要入口，每个连接体由中央大厅、大楼梯和步行廊组成，可贯通 4 个展厅。按照建筑物的功能性质，整组建筑由 4 个平行六面体展馆组成，色调统一为白色。几何形体的起伏、虚实和曲直对比，带拱顶的门廊、角窗、带圆弧的门楣等元素，简洁的门窗洞口和大片的实墙面，建筑技术与功能的完美结合，创造出具有强烈时代感的建筑造型。

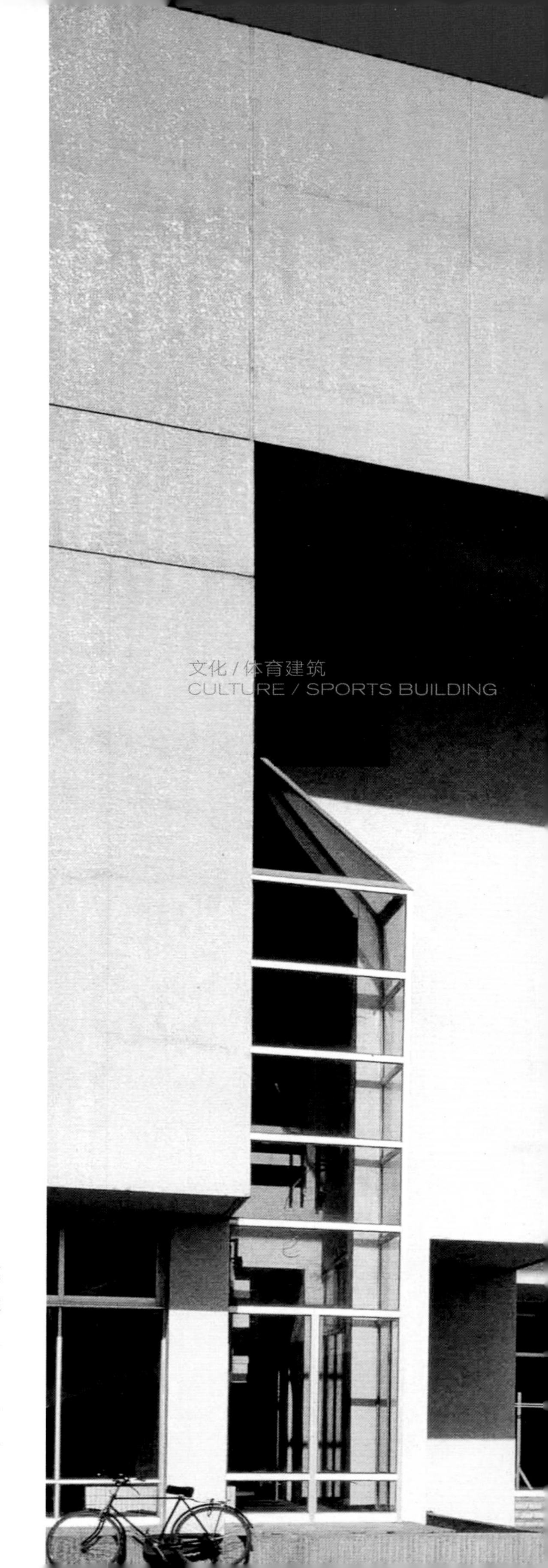

Hall No.2 to Hall No.5 are the first phase of the construction of the China International Exhibition Center. The four exhibition halls are connected by 3 connecting spaces. Considering the requirement of flexibility, each exhibition hall is defined as a 63mX63m square space. Inside the exhibition hall, a gallery corridor is pulled out of the enclosing walls at 4.5m height, which not only extended the space for exhibitions, but also diversified the spaces from both interior and exterior.

The four exhibition halls are connected by three spaces as the main entrance to the exhibition hall, each connecting space includes a central hall, a large staircase and a walking gallery that run through the four exhibition halls. The entire complex consists of four white parallel hexahedral structures. The undulating geometrical forms, the contrast between the imaginary and the reality, between curves and straight lines, arched porches, corner windows, the lintel with arc, the simple doors and windows as well as large area of solid walls have perfectly combined architectural techniques and functions, creating a form that strongly echoes the Zeitgeist of the era.

TRADE LIAISON OFFICE UPPER LEVEL BETWEEN
HALL 4,5

又见五台山剧院
Encore Mount Wutai

建成时间 Completion Year：2014
建设面积 Building Area：27 837m^2

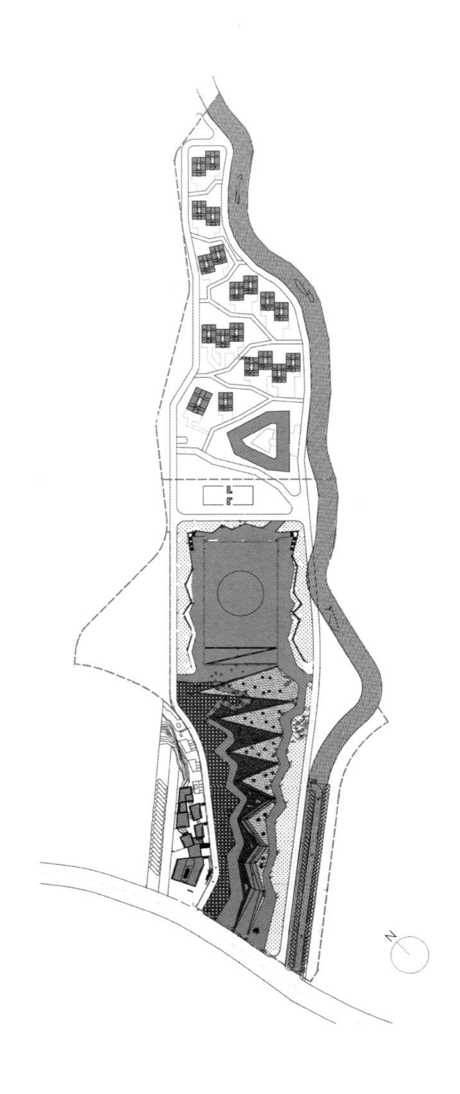

又见五台山剧院是导演和建筑师紧密合作的成果，使得表演和建筑在艺术层面融为一体。《又见五台山》剧目回应佛学主题，希望人们重新审视一生、一年、一天、一瞬间、一轮回，从中体味人生的价值，感知佛法的智慧。

剧场的主体是 72m×120m×24m 的矩形体量，为打破建筑固有的边界，从建筑主体向前广场延续出连续七折的墙体，宛若徐徐拉开的经折，巧妙地化解了大型建筑体量对于自然环境的压迫感，形成没有边际之感的建筑。经折墙共七折，取义“七级浮屠”，暗示人修行成佛的七个层次，每个折院中设主题预演区，将游客在观赏表演过程中，不知不觉引导至演艺中心主入口，游客的观演情绪达到高潮。

歎一切如來修諸功德種種苦行尸波羅蜜門或現不可說佛刹
微塵數剎截肢體羼提波羅蜜門或現不可說佛刹微塵

北京动物园大熊猫馆
The Panda House of the Beijing Zoo

建成时间 Completion Year：1990
建设面积 Building Area：1042m²

设计有机整合了新大熊猫馆和原熊猫馆，在旧馆和动物园北配楼之间设九宫格形铺地，铺地上设九根矮柱，为新大熊猫馆入口之暗示。铺地向东北斜向僻一直路，朝向新馆的正入口。沿斜路在接近熊猫馆正入口前布置两片平面为牛角形的竹林，作为熊猫馆入口的前奏。因熊猫喜阴，室外活动场地全部向北。四棵古树全部保留，并被巧妙地组织到圆形母题构图中去，形成定位感。室内环线为主要参观路线，观众可从多方位、多角度观赏珍品大熊猫的风采。

The design integrated the new panda house with the existing one. A pavage shaped like a nine-square pavilion was installed between the existing house and the zoo's north annex. Pavage were installed along the northeast/southwest line towards the main entrance of the new house. Two horn-shaped lots of bamboo were planted in front of the entrance to the new Panda House as a prelude. Since the pandas prefer shade, the outdoor spaces were all north facing. The design preserves the four old trees and subtly integrates them into the circle scheme to create a sense of positioning. The indoor loop is the main circulation for tourists, where visitors can enjoy a view to the pandas from multiple perspectives.

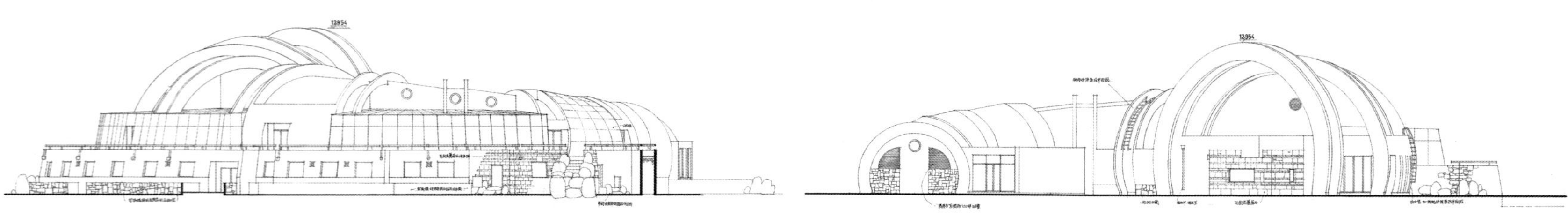

BOOKING OFFICE
ADMISSION TWO YUAN

珠海大剧院
Zhuhai Opera House

建成时间 Completion Year：2017
建设面积 Building Area：59 000m²

建筑以谦逊的姿态融入海洋与大地，成为自然界的组成部分。剧院包括 1550 座的歌剧院、550 座的多功能剧院和 350 座的室外剧场。源于自然的流动曲线沿海平面展开，勾勒出整体轮廓，显得轻盈而富有动感。用地布局取意于傍晚时分海潮退去、月光初起，海面与天空渐渐融合深邃的场景，形成“春江潮水连海平、海上明月共潮生”的古诗意境。从情侣路上、从伶仃洋上、从港珠澳大桥上，歌剧院以纯净优雅的自然形象融入大海和天空，如同被半透明的薄纱包覆，显得轻盈而迷幻，形成由环境到建筑的自然过渡。

This house is designed to blend the elements of the ocean and the earth in a humble way, to make it part of nature. The house consists of a 1,550-seat auditorium, a multipurpose space with 550 seats and an outdoor theatre with 350 seats. The architecture form originated from the natural flowing curve unfolding along the coast, showing lightness and dynamic power. The layout of the land is intended to create a poetic atmosphere where the sea and the sky mingle together and become one. Seen from the Lover's Road, the Lingdingyang estuary, and from the Hong Kong-Zhuhai-Macao Bridge, the opera house is integrated into the sea and the sky with its purity and elegance. The volume looks like it's covered with a translucent tulle, light and phantom-like, concluding the transition from the environment to the architecture.

HI CITY

中国科学技术馆
China Science and Technology Museum

建成时间 Completion Year：2008
建设面积 Building Area：102 280m²
合作公司 Partner：RTKL

中国科学技术馆与国家体育场、游泳馆等奥运场馆呈掎角之势，间隔大片开阔空间，在体量和形式感方面相互呼应，有助于强化奥林匹克运动中心区的整体空间和建筑意象。建筑采取单一的方形外观体量，意象来自儿童拼图游戏。若干积木般的体块单元穿插咬合，构成一个超尺度的巨型立体拼图。各“拼图单元”正面采用白色金属板，侧面为绿色反射玻璃，与“鸟巢”、“水立方”的红、蓝构成对比式的色彩关联，从而强化奥运公园的场所感和领域感。

The China Science and Technology Museum is located at sharp angels to the National Stadium, the Water Cube and other Olympic venues. The arrangement creates a large area of open space among them and echoing each other in terms of volume and form, which can help to strengthen the overall spatial and architectural image of the Olympic Sports Center. The building, in a single square volume, was designed in relation to a puzzle for children. A number of building-block-like bricks are interspersed to form a giant three-dimensional jigsaw puzzle. The “Jigsaw Unit” has a white metal plate on the front facade and a green reflective glass on the side, in contrast with the red of the “Bird's Nest” and the blue of the “Water Cube”, which further enhances the sense of place and the sense of an Olympic Park.

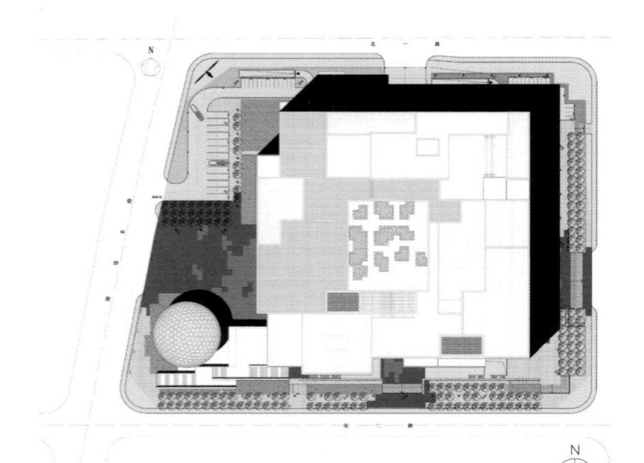

中国科学技术馆

中国电影博物馆
China National Film Museum

建成时间 Completion Year：2005
建设面积 Building Area：37 930m²
合作公司 Partner：RTKL

电影博物馆的建筑设计必须在文化性与娱乐性之间达到一种平衡。以一系列通俗并富娱乐性的基本建筑语素为出发点，通过艺术化的处理衍生出艺术和娱乐交融的严肃作品，如同一部影片的制作与欣赏过程。从最直接和易于领悟的外部视觉特征出发，吸引大众的注意力，由此引发好奇与期待心理，促使人们展开深层的探索。

建筑设计历经了一个对貌似单纯的题目进行深入解析挖掘，而最终将一系列复杂的内容提炼升华为一个简练而丰富的矛盾统一体的过程。

The architectural design of the film museum must reach a balance between culture and entertainment. Starting from a series of popular and entertaining basic architectural morphemes, the architects created a serious work from their fusion with artistic approach, which is similar to the production and appreciation of a film. In order to attract public attention, we started from exterior features that are most direct and easiest to understand, and further explored the design with stimulated curiosity and anticipation.

The design went through the process of in-depth analysis and mining for simple topics, resulting in a series of complicated elements that were refined into a contradictory unity that was concise and diversified.

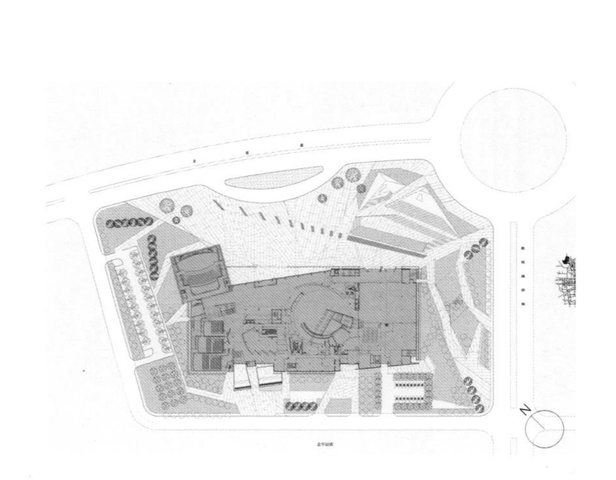

中国国际展览中心 I 期
China International Exhibition Center Phase I

建成时间 Completion Year：2008
建设面积 Building Area：180 000m^2
合作公司 Partner：TVS INTERNATIONAL

中国国际展览中心 I 期展馆布局模式兼具功能性和实用性。展厅使用可分可合，既可满足多个小型展览同时布展，互不干扰，又能将展厅相互联通，承办大型展会。便捷的交通、良好的流线是现代展馆高效运营的必要保证，展览中心人流、车流、货流清晰独立，互不干扰。中轴景观设计从功能出发，紧扣地域文化。内部庭院采用中国传统的院落式布局，富有民族特色，进一步体现出整体设计中所尊崇的人文精神。

The layout of the Phase I pavilion of the China International Exhibition Center is both functional and practical. The use of the exhibition halls is flexible, which meets the needs of multiple small exhibitions without disturbing each other, and also connect the exhibition halls to large-scale exhibitions. Convenient transportation and a reasonable circulation are necessary guarantees for the efficient operation of modern exhibition halls. The flow of people, vehicles and materials of the exhibition center are independent and clear without interference from each other. The central axis landscape design origins from basic functions and follows the local culture. The internal courtyard adopts the Chinese traditional courtyard layout, which expresses ethnic characteristics, and also embodies the humanistic spirit in the overall design.

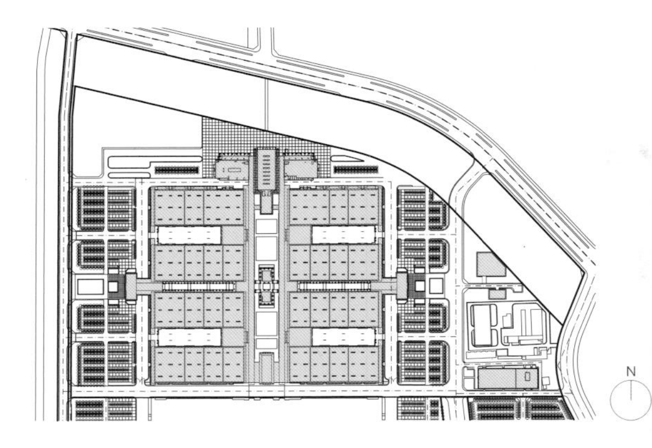

国家大剧院

The National Center for the Performing Arts of China

建成时间 Completion Year：2007
建设面积 Building Area：219 400m²
合作公司 Partner：ADPI

国家大剧院总体规划融合了水、绿色空间和人性化建筑三大主要元素。主体建筑是一幢由曲线构成的超大椭球壳体，通透的玻璃如同舞台拉开的幕布。当夜幕降临时，钛金属板上的蘑菇灯如同千万颗小星星，若隐若现，在绿树掩映中，有一种朦胧之美。

国家大剧院内设有歌剧厅、音乐厅、戏剧场及小剧场。四个观众厅根据不同的空间、预定的音质设计目标、追求的装饰风格，各自采用了独特的设计，使环境视觉效果与听觉效果得到了完美的融合。

The masterplan of the National Center for the Performing Arts of China integrates the three major elements of an architecture: water, green space and humanization. The main building is a large ellipsoid shell designed with smooth curves, with the transparent glass like an opening curtain of a stage. When night falls, the lights on the titanium plate are like thousands of little stars, hidden or visible in the shadow of green trees, creating an overall image of a hazy beauty.

The complex includes an opera house, a concert hall, a theatre and a smaller theatre. The four auditoriums are designed uniquely according to different spaces, predetermined sound quality, and decorative style to integrate the environmental visual effects and auditory effects perfectly.

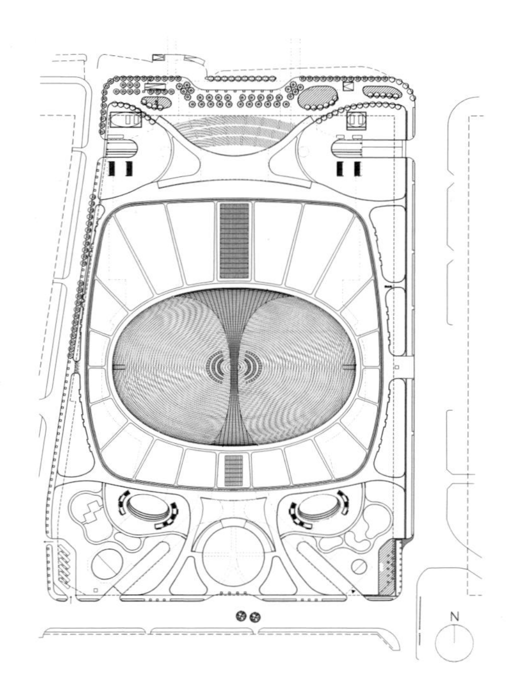

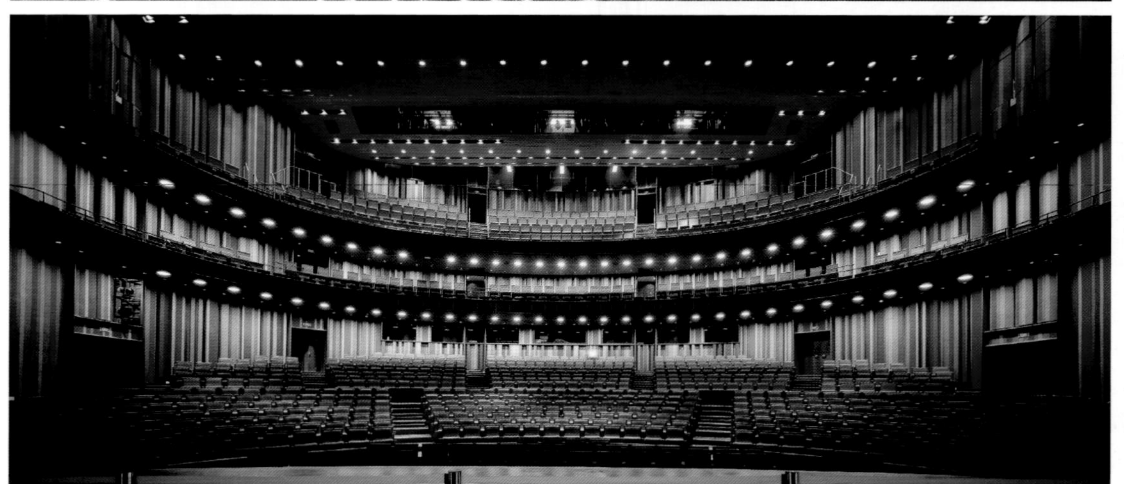

第九届中国（北京）国际园林博览会主展馆

Main Exhibition Hall of the 9th China (Beijing) International Garden Expo

建成时间 Completion Year：2012
建设面积 Building Area：54 044m²

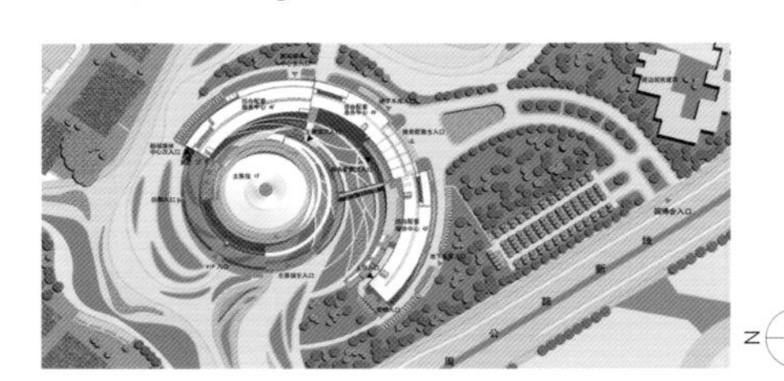

设计试图通过建筑形式与空间的语汇追寻生命之源的力量。建筑以主展厅为源起，以螺旋状生长、传播、辐射，最终融于园区的景观之中。

建筑由展厅及配套服务中心两部分功能组成。展会空间以一个直径为 70m 的中心展厅为核心，其他次展厅围绕此中心依次展开。中心展厅采用空间桁架结构，既可以满足特殊天气下大型室内演出的需要，又可以在展览期间根据不同的需求灵活划分。配套服务中心及其他附属空间在形态上则处理成由花蕊发散出花瓣一般的三段线性空间。

This design concept attempts to trace the power of life through the vocabulary of architecture and space. The building is based in the main exhibition hall, which grows, spreads and radiates in a spiral shape and which is finally integrated into the landscape of the park.

The building is composed of an exhibition hall and an ancillary service center that provides supporting function. The major element of the exhibition hall is a 70-meter diameter central hall, surrounded by other sub-exhibition halls. The space truss structure system of the central exhibition hall can meet the needs of large-scale indoor performances in bad weather, and can be adjusted flexibly according to different exhibition needs. The ancillary service center and other ancillary spaces are designed with three linear spaces like petals stretch froming the center flower space.

中国美术学院南山校区
Nanshan Campus of China Academy of Art

建成时间 Completion Year：2003
建设面积 Building Area：62 112m²

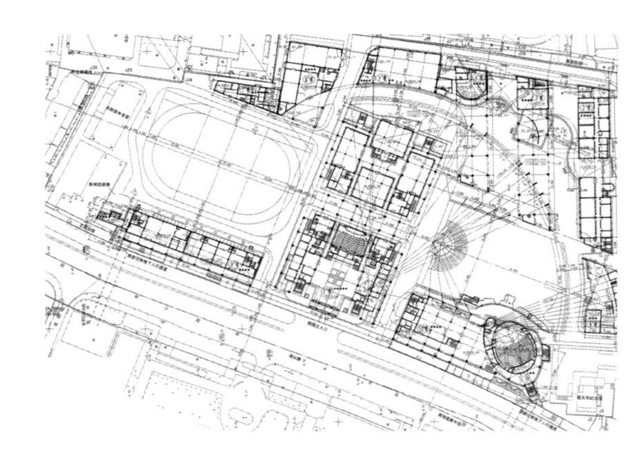

新校园建设项目主要包括教学教辅用房、各项配套设施及可对外开放的美术文化设施。用地内保留了潘天寿纪念馆、艺苑宾馆、教工宿舍食堂等建筑。校园设计因地制宜，在经营建筑群体开放空间格局和承续历史传统建筑特色方面进行了有意义的尝试。

体现至中意念：以中部主入口为校园中心，设复合型的核心建筑，交汇纵横方向、不同高度、不同层面的空间序列，构成纪念性文化广场的主题空间。惜重金角银边：建筑物沿用地周边布局，教学建筑均南北朝向，这对艺画尤为重要；沿街文化设施则直面社会，动静有别，互利互善，最大程度地形成校园中心的共享绿地和运动场地。核心建筑底部架空，视线纵贯校园南北。采用底层架空和江南建筑特有的高墙、深院、窄巷等处理方法，化解用地狭小、建筑密度过高所导致的空间拥堵感。以水墨韵味的黑白灰作为校园建筑色彩基调。在细部处理上，着意贴切地方特色。

中国美術學院

This project for a new campus mainly includes rooms for teaching aids, various ancillary facilities, and art and cultural facilities that can be opened to the public. Within the site, the Pan Tianshou Memorial Hall, Yiyuan Guest House, faculty and staff housing and canteen were preserved. The design was made to adapt local conditions, attempting to manage the open space of the architecture group with the effort to inheriting the traditional elements of local architecture features.

Focus on the campus center: It was decided that the campus center is where the the central entrance/exit located. A core building with composite functions was built on the site. This core building has intersecting horizontal and vertical directions, different heights, and different levels of spatial sequences, forming the theme space of the commemorative cultural square. Other buildings are arranged along the edges of the site, and the academic buildings are all north-south oriented, which is particularly important for painting spaces. The doors for cultural facilities were designed to face the street, insulating the noises and creating a quieter space in the campus. The high ceiling of the ground level of the core building creates the uninterrupted line of sight along the north-south axis. The high open space of the ground level, together with the high walls, deep courtyard and the narrow alley of the core building, features unique to Jiangnan architecture style, solves the feeling of congestion caused by the limited land and high building density. Black, white and grey like in ink paintings are the basic colour tones of campus architecture. Every detail of the design adapts to local context.

北京建筑大学 6 号综合服务楼
Service Building No. 6 of the Beijing University of Architecture

建成时间 Completion Year：2012
建设面积 Building Area：4443m^2

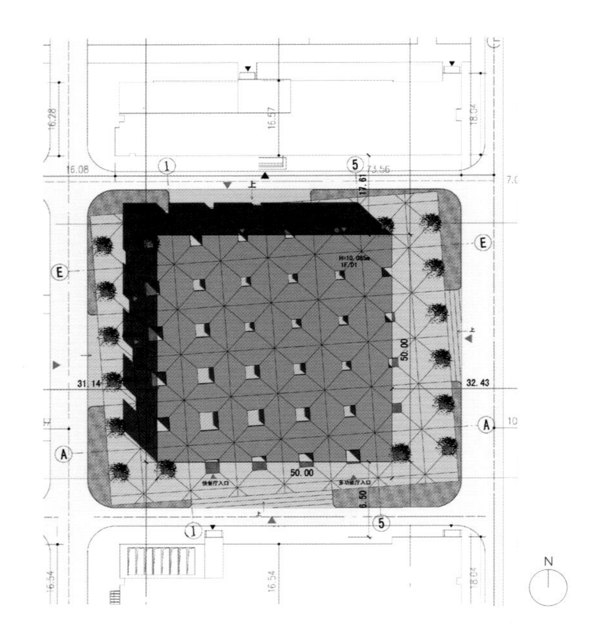

综合服务楼作为学生宿舍区内的一个小型公共建筑，功能的自由转换为设计构思的出发点。地上一层，无柱；采用单元式模块组合；电气、设备为功能转换提供有利条件；内部夹层采用较易拆除和可回收的材料；增加节能环保设施。

由10m×10m单元排列形成的60m×60m的正方形平面，每个单元由中间的一个天窗和四坡屋顶组成。在建筑外侧设置一层外廊，为外侧的大窗户提供遮阳，同时也为店铺和学生活动提供一个半室外空间。外廊的另一个重要功能是为大跨度的梁提供支撑。为丰富外立面，外廊在轴网旋转后将外侧单元进行切割处理。该建筑为钢筋砼结构，为了表达建筑的几何逻辑关系，屋顶、切割面为清水砼，切口内为木质外墙。

The service complex as a small-scale public building in the student dormitory area, the concept origins from the free transform of different functions. The one-floor building is modular combination mode without columns; electrical and equipment provide favourable conditions for function conversion. Removable and recyclable materials are used for the interlayer, also applies energy-saving facilities.

The 60m × 60m square plane formed by 10m × 10m units, and each unit consists of a skylight in the middle and a hip roof. A corridor is set outside the building to provide shading for the large windows outside, and also provides a grey space for the shops and students' activities. Another important function of the corridor is to provide support for long-span beams. After rotating the grid, the outer units are cut to make the facade multileveled. The building is steel-reinforced concrete structure, in order to express the logical relationship of architectural geometry, the roof and cutting edge surface are pure concrete, while the recessed part is decorated with wooden materials.

北川中学
Beichuan Middle School

建成时间 Completion Year：2010
建设面积 Building Area：129 242m²

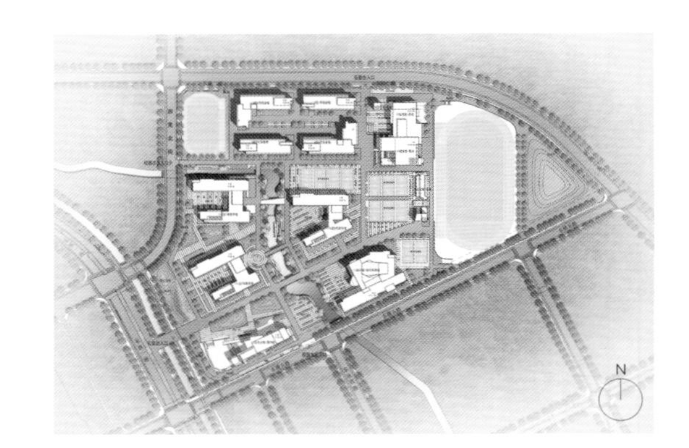

汶川地震灾后重建的新北川中学，设计团队从城市规划的层面思考校园规划，使校园总体规划和单体布局能够尊重和呼应北川新城的城市总体规划，并探索学校资源与社会（社区）共享的可能性，以及校园开阔空间在城市抗灾时作为避难场所利用的可能性。新北川中学的运营规模定位是 5200 名学生的完全寄宿制中学（超大规模），它既不同于一般规模、走读式的全日制中学，又在教学管理模式上不同于同等规模的大专与大学。我们亦针对超大规模尺度新建校园的特点，研究新北川中学学生日常教学管理和行为模式，规划中重点关注三个层面的问题：①定点时段不同分区地点集中、大量人流的迁移，简洁动线的满足；②普通课间时段各教学组团学生就地活动空间的消纳能力；③不同年级组团空间的层级划分和归属感的营造。

1/A
教学楼
嘉祥楼

The new Beichuan Middle School is a reconstruction project after the Wenchuan Earthquake. The design team targets on the campus planning from the perspective of urban planning, so that the planning and layout of the campus can respect and respond to the overall planning of Beichuan new town. At the same time, exploring the possibility of sharing between school resources and the society (community), and the possibility of using campus open space as a shelter for natural disasters. The operation scale of new Beichuan Middle School is a full boarding middle school with 5200 students, which belongs to the super-large scale. It is not only different from the general scale of full-time daytime school, but also different in teaching management mode from colleges and universities of the same scale. The design focused on the characteristics of super-large scale new campus, the designers did research on the daily teaching management and behaviour mode of students in the new Beichuan middle school, so the planning aimed at solving three questions of different aspects: the circulations design should meet the demand of concentrated and large number of people in a fixed time interval of different areas; the public space should meet the demand of in-situ activities for every teaching group during usual class hours; hierarchical division of grouping spaces for different grades and creating the sense of belonging at the same time.

望京科技园二期
Wangjing Science and Technology Park Phase II

建成时间 Completion Year：2004
建设面积 Building Area：46 297m²

望京科技园二期是一个供高科技企业使用、配套齐全的办公建筑。该工程由三幢平面类似的建筑和一个连接体组成。两个主体量分别体现了矩形沿折线轨迹翻转的感觉，另一个体量是由密肋玻璃与全透明幕墙组成的从半透明到全透明的体量。主体块外墙采用低辐射印刷玻璃，经光线折射造成一种独特的图案效果。建筑布局采用占边的方法，以便形成秩序，同时在一期的北侧形成广场。景观绿化贯穿整个用地，并使二期与一期之间形成良好的关系。

Phase II of Wangjing Science and Technology Park is a complement office building with comprehensive facilities for high-tech enterprises. This project consists of three buildings with similar layouts and a connection space. The two main volumes reflect the image of a rectangle turning along a folding line track, with the other volume made from semi-transparent to fully transparent materials, which is composed of ribbed glass and a fully transparent curtain wall. Low-radiation printing glass is adopted for the facade wall of the main building, and the refraction of light creates a unique pattern effect. The layout of the three buildings is set along the edges of the site to form a strong order and create a square on the north of the phase I project. Landscapes run through the whole land and generate a good relationship between phase II and phase I.

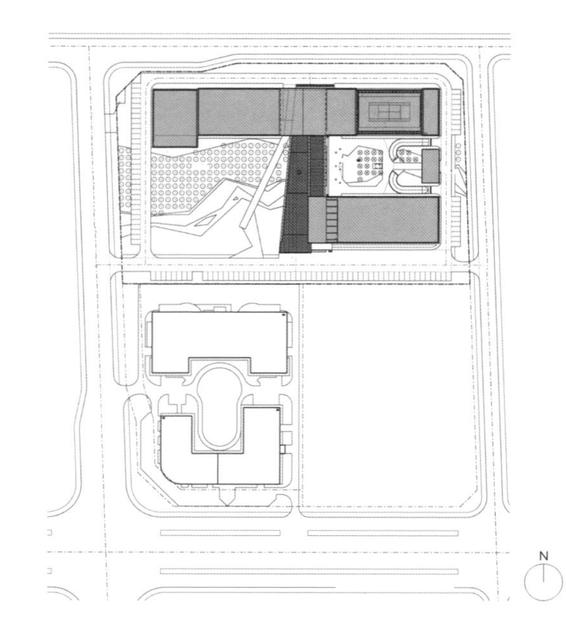

联想园区 C 座
Block C of Lenovo Park

建成时间 Completion Year：2004
建设面积 Building Area：94 884m^2
合作公司 Partner：SOM

联想园区 C 座采用双板连体建筑外形，建筑立面以轻盈通透的玻璃为主，配以灰色石材、金属遮阳板。全玻璃大堂采用悬索结构，晶莹剔透、雄伟恢弘。办公楼东侧三层以上采用了斜拉索，向东悬挑 9m，形成三面采光、开敞通透的办公环境。大楼西侧及十七层采用双层呼吸幕墙，南侧采用水平丝网印刷玻璃遮阳板，用来遮挡直射光及避免眩光。底层双层挑高的外廊与上部落下的片墙、过廊形成丰富的流动空间，行走其中可感受到中国传统园林的意境。

The Lenovo Park C Block adopts the shape of the double-panel building. The facade of the building is mainly made of light and transparent glass, with grey stone and metal sun visors. The glass lobby features with a suspension structure which is crystal clear, forming a majestic space feeling. The stay cables are used for the floors above the 3rd floor on the east side of the office building and cantilevered 9 meters to the east to form an office environment with three sides of light, open and bright. The double-layer respirable curtain wall is used on the west side and the 17th floor of the building, and the horizontal screen printing glass sun visor is used on the south side to block direct sunlight and avoid glare. The two-story overhead space with the walls going down and the corridor together form a flowing space, where people can feel the artistic conception of Chinese traditional gardens while walking.

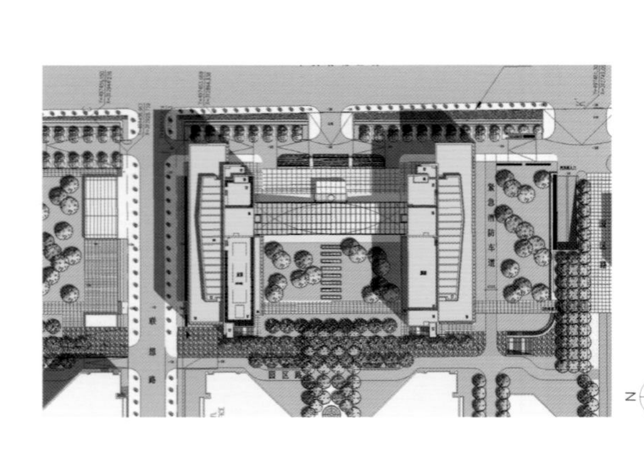

联想研发基地
Lenovo Research and Development Base

建成时间 Completion Year：2004
建设面积 Building Area：96 156m^2

联想研发基地沿规划道路将建筑单体分为东西南北四组，顺应用地轮廓将外围控制紧密，取整体连贯线性排列，南北两组为研发用房，东西为点式单体对景布局，以流水贯通，并用弧廊将东、西、南、北楼第三层相连，设参观专用通道。建筑组团对外连续围合，对内是轻松布局的草坡、流水、石桥、瀑布等园林景观。此项目中，清水混凝土不仅在国内首次得到大面积使用，而且同时呈现出多种表现形式，是本项目最大的特色，也最具挑战性。

The design of Lenovo Research and Development Base divides the whole building volume into four clusters according to the four directions along the planned road. The project complies with the contours of the site to control the periphery so as to form a coherent linear arrangement. The north and south clusters are research spaces, and the east and west are single-frame layouts with point-type, and beautiful waterscape running through it. The arc corridor is set to connect the third floor of the four building clusters with a specific visiting passageway. This group of buildings is continuously enclosed, while the inside space is set with natural landscapes such as grass slopes, flowing water, stone bridges, and waterfalls. Another important point of this project lies in that pure concrete is used for the first time in China, which also presents a variety of forms. This is not only the outstanding feature of this project, but also the biggest challenge.

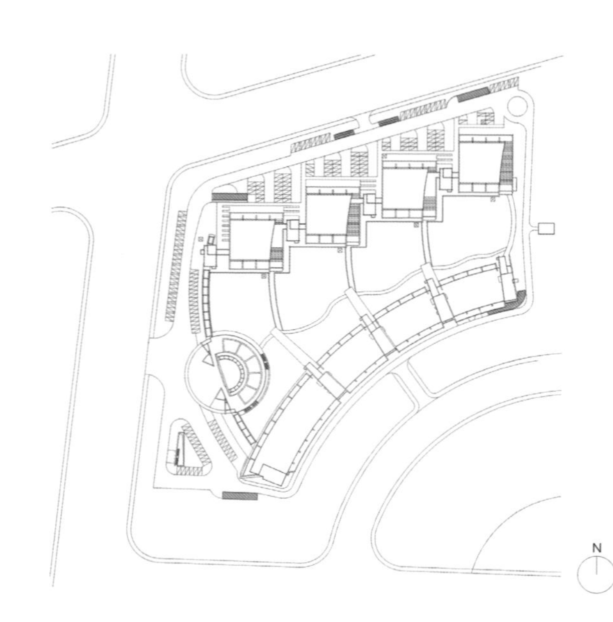

腾讯北京总部
Tencent Beijing Headquarters

建成时间 Completion Year：2019
建设面积 Building Area：334 386m^2
合作公司 Partner：OMA

腾讯北京总部位于北京市上地中关村软件园二期，由两栋独立的建筑组成，其中主楼可容纳约 6500 名员工，主要功能为科研办公和配套服务设施。主楼东侧为能源中心配楼，为主楼提供能源支持。约 33 万 m^2 的办公功能整合在主楼中，在单层 28 000m^2 的超大平面中，为了解决核心区的采光与通风问题，均匀设置了 6 个庭院：2 个室外庭院，4 个室内庭院。标准层通过环路、斜街、内街划分开敞的办公区，形成相对独立的 12 个办公单元。

Tencent Beijing Headquarters is located in phase II of Zhongguancun Software Park, Shangdi District. The main building can accommodate about 6,500 employees, and the main functions are scientific research offices and ancillary service facilities. The east side of the main building is the ancillary building which major function is to provide the necessary energy for the main building. About 330,000-square-meters of offices are integrated into the most important building. In the large-scale plane which has 28,000 square meters per floor, in order to solve the lighting and ventilation problems in the core area, six courtyards are evenly arranged, including two outdoor courtyards and four indoor courtyards. On the standard floor, the open office area is divided through loops, oblique streets, and inner streets to form a relatively independent 12 office units.

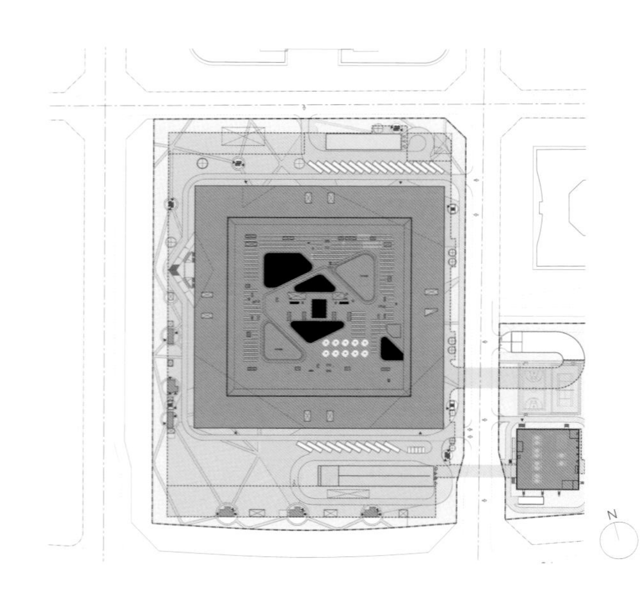

停车场

小米科技北京总部
Xiaomi Technology Headquarters Beijing

建成时间 Completion Year：2019
建设面积 Building Area：346 686m²

园区不规则的平行四边形用地被巧妙利用，八个标准办公单元体构成三栋建筑，纵横交错地坐落在场地上。适合北京气候特征的正南北向布局编织出一张建筑逻辑网。逻辑网延续到地面，在景观和广场道路设计上进行扩展和控制，与用地边界自然交汇。交汇后的场地及室外空间以下沉庭院为中心，渗透到各个单体的底层，形成连续而有变化的诸多室外环境。建筑的外表皮由单元式幕墙构成，每个单元跨越两层层高，整体形象简洁而挺拔。布纹不锈钢钢板内框具有细腻朦胧的反射效果，令人感受到细节的精致。

This irregular parallelogram land in the park is used efficiently, with the three buildings consisting of eight standard office units, which crisscross the site. The buildings are north-south oriented, which is reasonable for the climate features of Beijing, forming a logic framework for architecture. The logic framework extends to the ground and controls the landscape design, as well as controlling the square and roads that naturally connect to the edges. The site and the outside space are centered with a sinking courtyard, which penetrates into the bottom layer of each unit which form a continuous and varied outdoor environment. The facade of the building is composed of unitized curtain walls, and each unit spans two floors, making the overall image modern and forceful. The inner frame of the cloth-pattern stainless steel plate has a delicate and sinuous reflection effect, making the details more exquisite.

低碳能源研究所及神华技术创新基地

Low Carbon Energy Research Institute and Shenhua Technology Innovation Base

建成时间 Completion Year：2014
建设面积 Building Area：325 354m²

以位于场地中央的神华学院为中轴，以低碳能源研究所和神华研究院为两翼，形成“一轴两翼”的规划骨架结构。以“浮岛”为建筑的外部空间意象，以“岩石”为造型意象。神华学院位于场地中央，以南北方向的主轴线组织整个建筑群体空间，以主入口的“大门洞”造型为起点，以 60m 高的职工宿舍楼为轴线的终点，依次布置会议中心、教学楼、职工宿舍及配套、职工健身楼等建筑单体。建筑按功能组合成两个相对完整的内向型庭院。

Taking Shenhua College as the central axis, which is located in the center of the venue, the Low Carbon Energy Research Institute and Shenhua Research Institute are set as two wings on both sides, forming the planning structure called “one main axis with two wings”. The “Floating Island” is the external space image, and “rock” is the architectural image. The Shenhua College is located in the center of the site, organizing the space of the entire building group, with the main axis running from north to south. The axis begins with the “Grand Door” and ends at a 60-meter high staff dormitory. Along this axis, there is a conference center, research and teaching building, staff dormitory with ancillary structures, employee fitness building and other units. The buildings are functionally combined into two relatively complete inward courtyards.

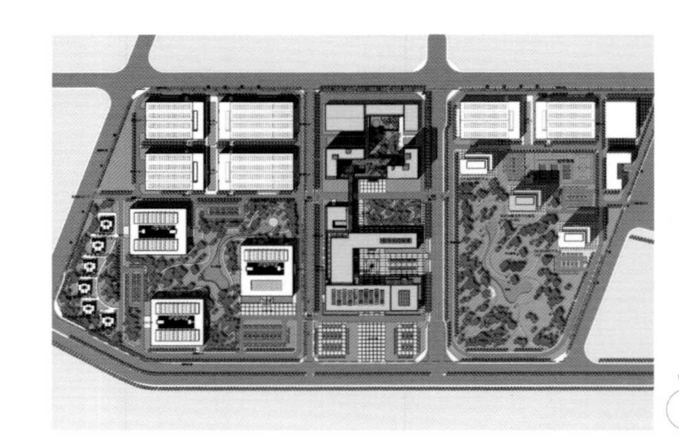

国电新能源技术研究院
Guodian New Energy Technology Research Institute

建成时间 Completion Year：2014
建设面积 Building Area：243 100m^2

项目建设内容包括：八个研究所的研发楼、实验楼和试验厂房等用房，以及科研办公楼、会议中心等配套设施等。整个园区分为东、西两个部分，东侧为研发实验区，是整个园区的核心；西侧为配套附属设施。通过研究各入驻单位研发活动的工艺流程和使用者的工作模式，创造一系列丰富的室内外建筑空间。主体建筑组群围合成一个矩形庭院，建筑单元高低错落，形成的屋顶平台为工作人员提供了更多交流、休息空间，营造出安静内向的研发环境。

The project includes eight research and development (R&D) buildings which belong to eight research institutes, laboratory buildings, experimental plants and affiliated facilities, along with research offices, a conference center and other ancillary facilities. The overall plan of the park is divided into two parts, east and west. The research and development experimental area situated on the east side is the core of the park. The affiliated facilities are situated on the west side. Based on the study of different institutes with different research processes, research and development activities and the users' working patterns, the designers created a series of creative outdoor public spaces. The building cluster is enclosed and forms a rectangular courtyard. The building units with different heights provide roof platforms for communication and rest for the staff, also creating a quiet and inward research environment.

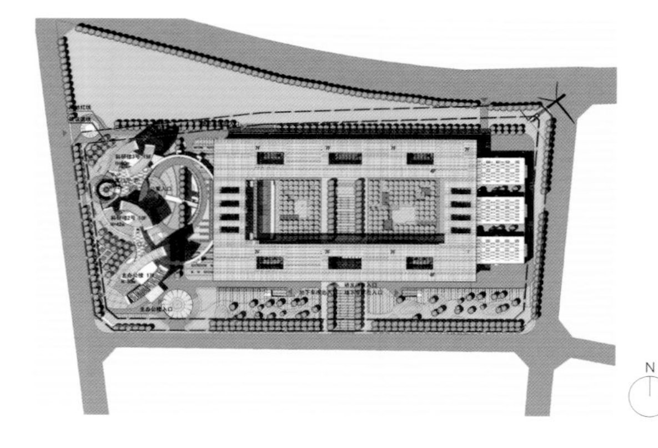

中国国电

中国石油大厦
China Petroleum Building

建成时间 Completion Year：2008
建设面积 Building Area：200 838m^2
合作公司 Partner：TFP

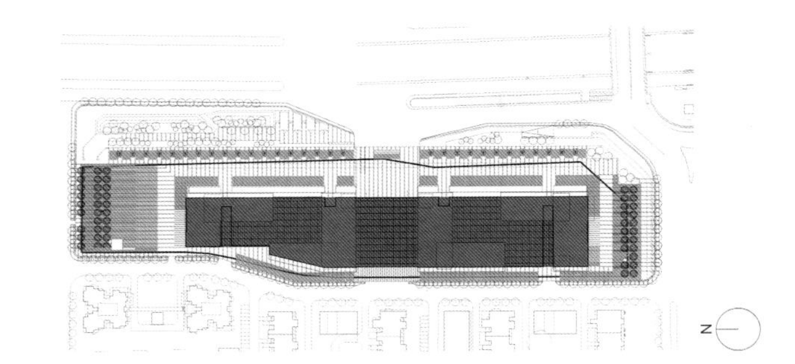

中国石油大厦位于北京市东城区东直门桥西北部。功能包括办公、会议、展览、商务、数据中心，以及档案、食堂、健身活动中心等附属服务设施。

集群组化的协调单元：为化解南北 347m 狭长用地的不利条件，选定 4 个 L 形的协调单元衍生建筑集群，创造性地解决了 300m 街墙的设计问题，提供了东西两条街道的景观联系，最大限度地提供南北朝向的办公室。一体两翼的空间格局：北部集群为集团公司和文化广场，南部集群为股份公司企业和广场，中部为总部核心区主中庭。长达 260m 的公共空间贯穿基地的南北两大集群，体现了总部的凝聚力和超人的魄力。办公建筑的中庭系统：中庭空间营造高效、舒适、纯净的办公氛围，高标准的人均立体绿化，提供了内部和谐的生态环境。弹性开放的模块模数：办公空间南北 8m 浅进深，东西 16m，三面采光。自由分割的灵活布局适应不同部门的使用需求。

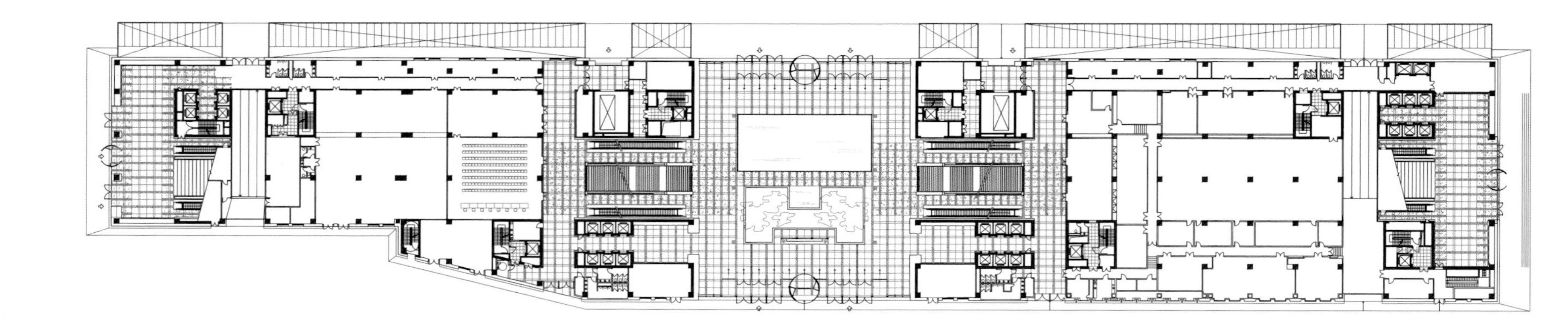

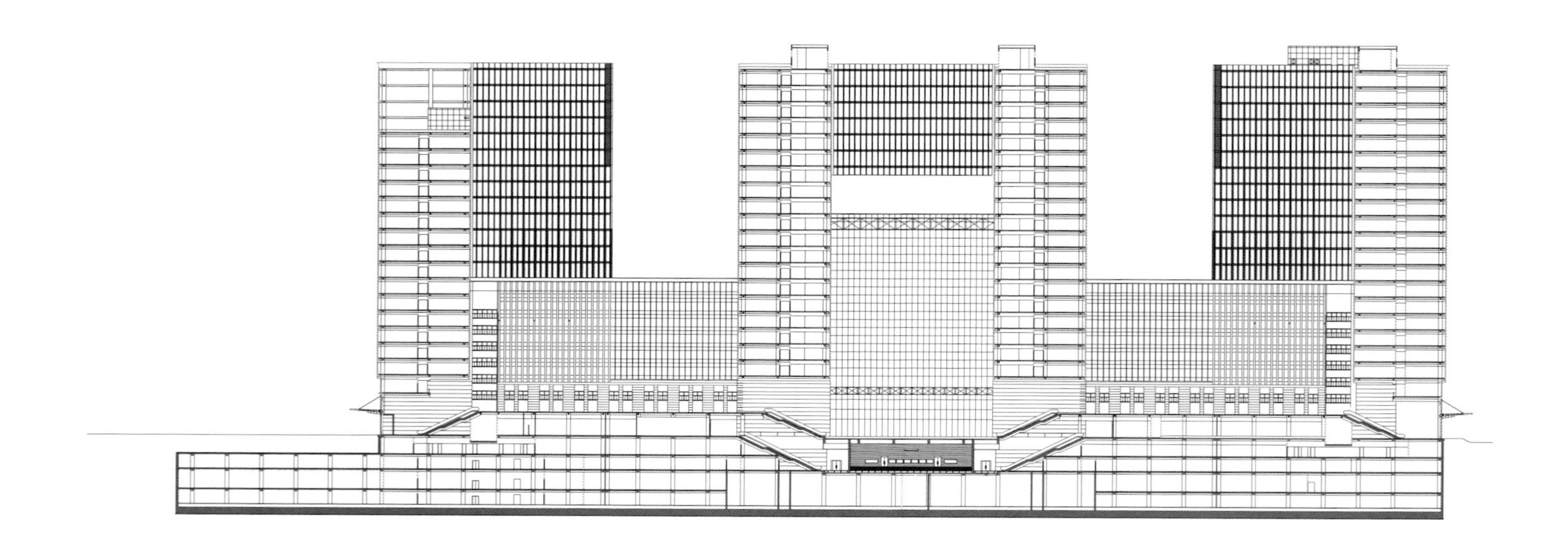

Located to the northwest of Dongzhimen Bridge, in Dongcheng District, Beijing, the China Petroleum Building hosts the functions of office, conference center, exhibition hall, business, data center, as well as archives, canteen, fitness center and other ancillary service facilities.

The coordination unit of the cluster organization: in order to solve the unfavorable land shape which is 347-meter long and narrow from north to south, four "L" shaped coordination units are selected to create building clusters. In this way, the 300-meter long street wall can be designed creatively, providing a connection for two streets on the east and the west. The layout maximizes the north-south orientation rooms for the office. The planning of the clusters is centered on a central courtyard in the core area. The north cluster includes the main building and culture square, while the south cluster includes joint-stock companies and a square. The 260-meter long public space runs through the two clusters from north to south on the ground floor, reflecting the cohesion and courage of the headquarters. The courtyard design: the atrium space has an efficient, comfortable and pure office atmosphere, while the high standard greening provides a harmonious and ecological internal environment. Elastic and open module design: the module of the office room is 8 meters deep and 16 meters wide, with lighting on three sides. This flexible layout, which can be divided freely, meets the needs of different departments.

侨福芳草地
Parkview Green

建成时间 Completion Year：2010
建设面积 Building Area：200 838m^2
合作公司 Partner：综汇建筑设计事务所
Integrated Design Associates

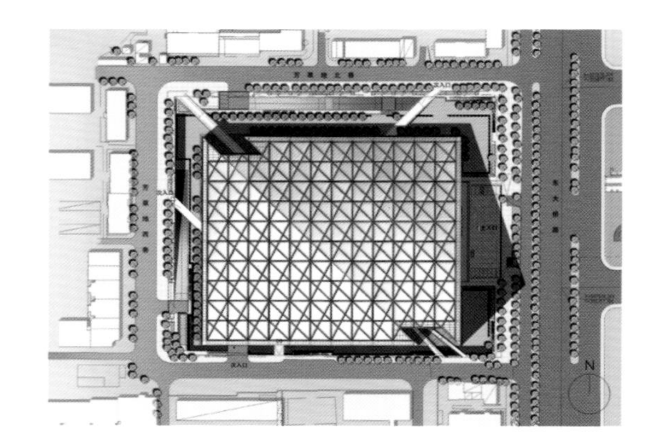

侨福芳草地是一个集商场、办公、酒店为一体的综合商务楼。建筑由 2 高 2 矮 4 个相对独立的单元组成，坐落在一个 10m 深的下沉花园上。建筑底层相互连通，上部由一个巨大的、能自由呼吸的环境保护罩包覆，能够自动应对不断变化的天气、温度、太阳角度、湿度和风向。环保罩高度超过 80m，立面采用单层玻璃，屋顶采用 ETFE 半透明膜。为了满足内部建筑高度的要求，环保罩形状为三角形，屋顶沿西面和北面倾斜。

每栋建筑都巧妙地与室外花园平台贯通起来，使整座建筑实现了自然通风。下沉区域为综合商业中心，由自然采光的中庭环绕。商场中有一座跨度 225m 的步行桥对角穿越建筑，使人们可以更加便捷地行走在 4 个建筑之间。建筑三至十二层为商务办公区，可以从沿街的朝向或中庭获得充裕的自然光。而且，这里 50% 以上的办公区可以直通空中花园、景观桥或露台。高塔 A、B 栋十三至十八层为个性化精品酒店。玻璃和钢结构搭建的空中大厅位于整栋大楼的最高点，可将城市美景尽揽眼中。

The Parkview Green project is a commercial complex that includes shopping malls, offices and a hotel. Located on a 10-meter deep sunken garden, the project consists of four comparatively independent units, two higher volumes and two lower volumes. The ground levels of the buildings are connected to each other, and the upper parts are covered by a huge environmental protection shelter, which adjusts itself automatically in response to the changing weather, temperature, solar angles, humidity and wind direction. The height of the cover is over 80 meters and the facade adopts a single-layer glass, while the roof adopts an ETFE translucent membrane. In order to meet the requirements of internal building height, the environmental protection cover is triangularin shape, and the roof is inclined along the west and north sides.

In the Parkview Green, each building is connected with the outdoor platform, which realizes the natural ventilation for the whole building. The sunken area is a commercial center surrounded by courtyards with natural lighting. There is a pedestrian bridge with a span of 225 meters which crosses the building diagonally so that people can walk between the four buildings conveniently. From the 3rd to the 12th floors are business offices, which can obtain abundant natural light directly. Moreover, more than 50 percent of the offices connect to the sky garden, landscape bridge or terrace directly. The 13th to the 18th floors of tower A and B are used as a boutique hotel. The suspending hall with glass and steel structures is located at the highest point of the whole building, where people can enjoy a panoramic view of the city.

ALFRED DUNHILL
Great

北京城市副中心行政办公区

Administrative Offices of the Beijing City Sub-Center

建成时间 Completion Year：2018
建设面积 Building Area：600 000m^2

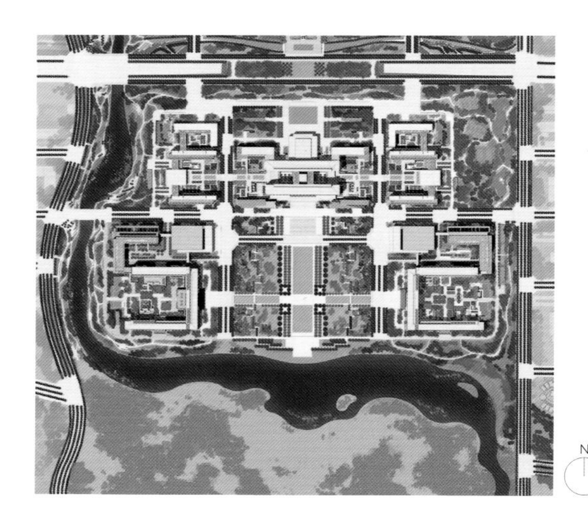

北京城市副中心行政办公区承担的政务中心功能是疏解中心城区非首都功能的重要标志性工作，使其成为北京城市副中心的重要功能组团。

中轴线序列空间突出整体建筑环境的秩序感；院落式排布的序列空间作为城市副中心建筑群排布的主干，向南北延伸，突出了政府端庄稳重的形象。建筑群呈品字形围合布局，并形成东西轴线，进一步突出了政府办公区的礼仪风貌。高低错落的建筑形态分区明确，使用灵活。建筑形象突显中式建筑庄严、典雅的特征，将传统建筑大屋顶曲面进行简化，并与当代设计理念和建造技术相结合。镜河自西北向东南环绕建筑群，在南部形成镜湖，与千年守望林起伏的地形相呼应，塑造丰富的自然景观效果。街坊式规划单元适应城市的可持续发展，强调小街区的开发理念，促进交通功能和公共服务功能的有机渗透。

The Beijing City Sub-Center undertakes the functions of the government affairs center, relieves the important works of the non-capital functions of the central city. It is an important functional building group of the Beijing City Sub-Center.

The central axis highlights the space sequence and strong order of the overall environment. The courtyard layout as sequence space, which is the main organization of the center building group, extends from north to south, highlighting a dignified and modest image of the government. The building group is designed in the shape of a Chinese character “ 品 ”, and forms an axis from east to west, highlighting the etiquette that fits to government offices. The buildings with different heights have clear functional divisions and flexible forms. The architectural image highlights the solemn and elegant features of traditional Chinese architecture.The grand curved roofs, a prominent feature of Chinese architecture, are made sleek and simple, and combines them with a modern style and construction technology. The Jing river goes through this area moving from northwest to southeast, forming Jing Lake in the south, which corresponds with the forest and mountainous terrain, creating a rich natural landscape effect. The neighborhood-style planning unit adapts to the sustainable development of the city, emphasizing the concept of small blocks while promoting the traffic and public service functions.

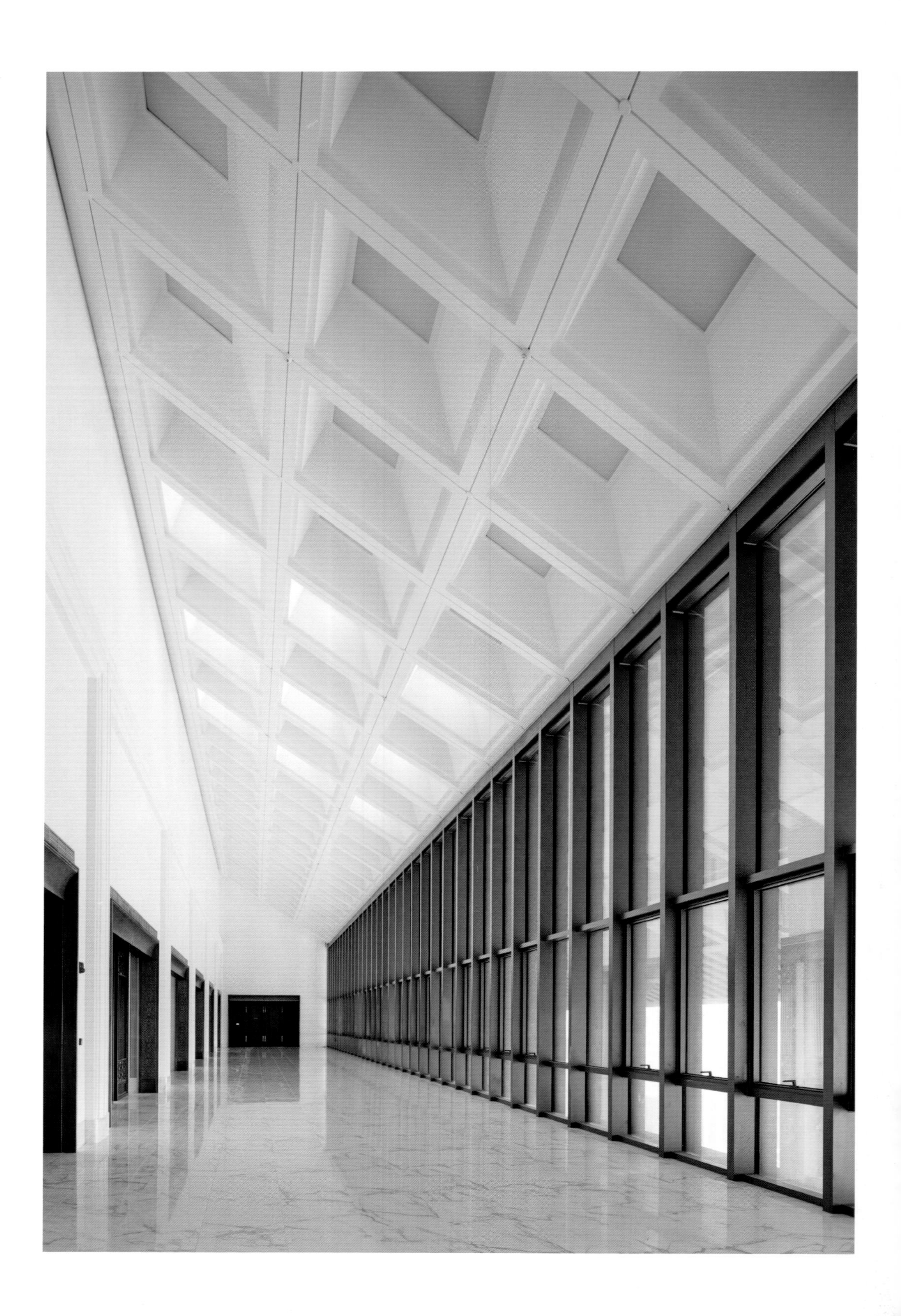

北京奥体商务南区
OS-10B 城奥大厦
South Olympic Business District
OS-10B Cheng'ao Building

建成时间 Completion Year：2018
建设面积 Building Area：70 000m²

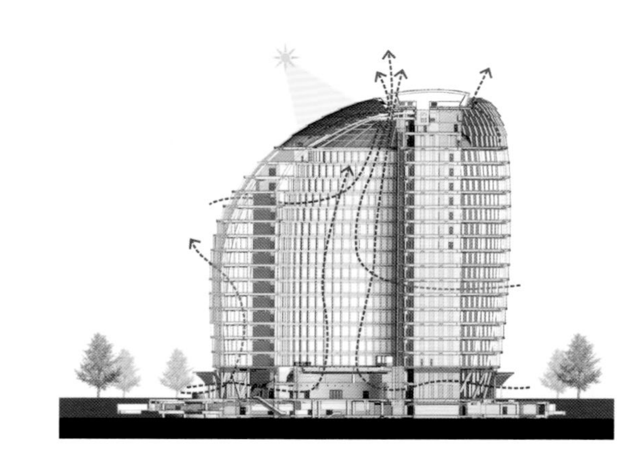

奥体南区 OS-10B 大厦作为非标准大型复杂性公共建筑，全面借助数字技术进行设计及建造控制。项目位于奥体文化商务园区中心景观 10 号地的东北角，东至规划奥体南区四号路，南及西至 10 号地规划集中绿地，北至规划奥体南区 2 号路。规划总用地面积 12 273m²。主要功能为商务办公，地上 19 层，地下 3 层（已建成），建筑最高点 95m。

OS-10B 大厦基于场所环境、使用者感受、技术性能等多方诉求，在建筑的生成和深化过程中制定了严密的控制逻辑。例如圆润倾斜的球状形体实际是以周围建筑的视线和气流因素作为控制参数不断拓扑变化而成的。同时这一形体将体形系数控制到最小，更有利于节能。它既有基于参数化、仿真模拟的高精度特征，同时又是一个自由且“个性化”的不可复制的定制设计成果——凸显其在几何控制体系，定制的钢结构体系，拟合自由形体的幕墙体系，幕墙与土建、结构系统的精细化协同，基于性能目标的建筑整体设计策略等方面的先进性。各参建方目前还在尝试以设计团队的数字模型作为中心基础模型，实现全生命周期 BIM 应用，以达到“数字孪生”的目标，让 OS-10B 大厦真正映射出建筑的未来。

The OS-10B Cheng'ao Building in the South Olympic District is a non-standard complex building, which is designed and constructed with the help of digital technology. The project is located in the northeast corner of the No.10 central landscape area inside the Olympic Culture and Business Park. On the east is the No. 4 Road, with the south and west enclosed by a collective green space, and the No. 2 Road sits on the edge of the north. The planned land area is 1.2273 hectares, and the overall gross floor area above the ground is 69,929m^2 (the underground part has been completed). The main function is business offices, with 19 above-ground floors and 3 underground floors; the highest point of the building is 95 meters high.

The OS-10B Building is based on meeting multiple demands of the environment, the users and technical performance, so a strict logic was established to control the design process. For example, the round, inclined spherical shape is formed taking into account people's sightlines and taking air current factors from surroundings as the control parameters. At the same time, the volume coefficient is controlled to the minimum value which can realize energy saving. It not only has the characteristics of high precision based on parameterization and simulation but is also a free and unique customized design that cannot be replicated, highlighting its geometric control system, customized steel structure system, and curtain wall system. With a fine coordination of curtain wall, civil engineering, and structural system, the application of advanced technologies is reflected in design strategies which are based on performance objectives. The participating partners are still trying to implement the full-life BIM application with the digital model as the foundation model, to achieve the goal of "digital twins". The OS-10B Building truly reflects a future architecture design.

全国人大机关办公楼
Administrative Office of the National People's Congress(NPC)

建成时间 Completion Year：2010
建设面积 Building Area：83 000m^2

全国人大机关办公楼是全国人民代表大会常设机构的办公场所。方案采用简洁有力的城池形轮廓，使原有建筑和新建筑围合成一个完整的整体，在体量上与天安门广场和人民大会堂相协调，同时对南广场西侧松散的建筑群起到整理作用。

项目借鉴北京传统四合院的特点，从功能出发，采用庭院式布局，以现代办公楼的尺度为基准，衍生出一系列内向的、独立的庭院。每个庭院都可以作为独立的办公单元来使用，以适应多部门、多层次的办公划分要求。根据北京的气候条件，选择绿色技术策略，并进行优化整合。因此其是第一个获得绿色三星认证的政府办公建筑，成为全国政府办公建筑中绿色、节能的典范。

The Administrative Office is the permanent establishment of the National People's Congress (NPC). The design adopts a simple but powerful ancient-city plan to integrate the new buildings with existing structures. The volume of the building is determined in comparison with Tian'anmen Square and the Great Hall of People, and also compensates the loose order of the buildings on the south square.

This project draws on the characteristics of traditional courtyards in Beijing, focusing on the distribution of the functions and adopting the courtyard layout. Based on the scale of modern office buildings, a series of inward and independent courtyards have been derived. Each courtyard can be isolated as an independent office to meet the demands of different departments and the division of multi-level offices. According to the climate conditions of Beijing, green technologies were applied and optimized. The new building has been awarded "Three-stars Green Certification", making it an energy-saving green building model for government office buildings.

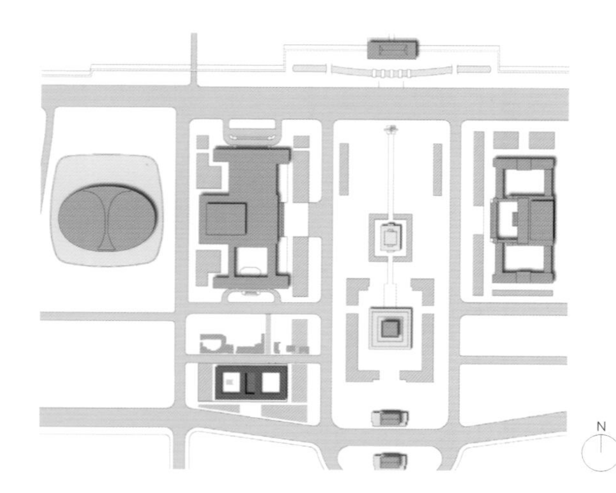

商务部办公楼改造
Ministry of Commerce Office Building Renovation

建成时间 Completion Year：2005
建设面积 Building Area：65 760m^2

中国商务部办公大楼位于长安街南侧，是构成东长安街城市天际线的重要组成部分。原大楼主体由三座板式高层建筑组成，此次对内部功能布局及外部形态进行了全方位改造。将中国传统建筑的三段式构图与西方古典建筑的比例进行融合，创造出整体感和秩序性，并将原建筑南北向和东西向的错台全部填平补齐。长约 260m 的建筑主立面只采用了一种开窗方式，在方窗间夹了 250mm 厚的石材带，玻璃退后的处理增强了石材的厚度感和体量感。为了使面对长安街的形象统一对称，屋顶只能做成披檐式坡屋顶，柱子演化成南北纵深的墙。

Located in the south of Chang'an Street, the Ministry of Commerce Office Building is an important component of the city skyline of Chang'an Street. The original main building is composed of three high-rise volumes,

and the renovation design transforms both internal functions and external form. Combining a three-stage composition of traditional Chinese architecture with Western classical architectural proportion creates a sense of unity and strong order. Meanwhile, the original staggered platforms in the north-south and east-west are made level. The main facade of the building, the length of which is about 260 meters, employs only one window pattern, with a 250-millimetres thick stone belt in between the square windows. The concave glass enhances the feeling of heaviness and volume of stone. In order to coordinate to the surrounding buildings on Chang'an Street, sloping roof and the south-north wall evolved from pillars are utilized in the design.

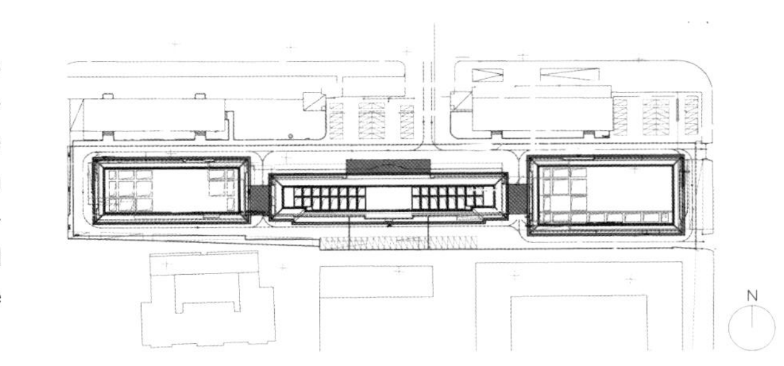

北京国际金融大厦
Beijing International Finance Building

建成时间 Completion Year：1995
建设面积 Building Area：100 000m^2

北京国际金融大厦由北侧 4 栋相对独立的办公楼和南侧两个弧形连接体组成。北侧办公楼为 11 层，首层为银行营业厅，二层以上为写字间。南侧楼为 14 层。大厦地下有两层。4 个长方形办公楼的底层分别是 4 个银行营业厅，它们由位于大厦中心的一个圆形大厅组织在一起，满足商业需求的同时，也有利于业务管理。大厅内部高 10m，中央有一个钻石锥顶。4 个塔楼成为该组建筑的标志，塔楼间有两个巨大的门洞和弧形连接体，使人们在有限的中央大厅内感到空间无限。

The Beijing International Finance Building is composed of four relatively independent office buildings on the north, and two connected volumes in the shape of arc on the south. Each of the tower-shaped north buildings has 11 floors; the first floor serves the function to provide services to the banking clients, the second floor and above are offices. Each of the south buildings has 14 floors above ground and two floors underground. The four banking halls on the ground floor of the four rectangular office buildings are arranged to circle around a circular-shaped hall in the center. The design can facilitate not only the efficiency of banking business but is also conducive to efficient management. The interior height of the circular hall is 10 meters, with a diamond cone in the center. The four towers have become the symbol of this building cluster, with two huge doors and arc-shaped connections between the towers, endowing the limited central hall with infinite feelings.

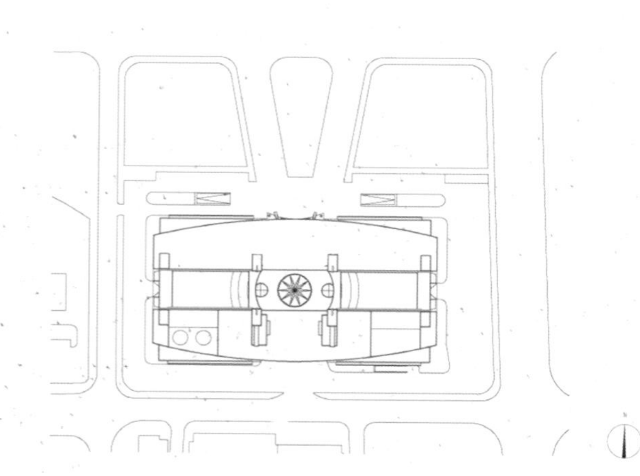

北京市高级人民法院
Beijing Higher People's Court

建成时间 Completion Year：2005
建设面积 Building Area：47 430m²

北京市高级人民法院包括审判楼、立案信访楼、安检厅三个单体，包含一个 2000 m²的共享大厅、15 个法庭、一个小型法律图书馆、1000 m²的立案和信访等法律服务空间，以及其他审判配套设施。

为实现空间的公共性和社会服务的便利性，审判楼面向东二环展开，临东二环留出开阔的城市广场。来访者可以从西入口广场，沿 70m 宽、4.5m 高的台阶拾级而上，通过一个 30m 宽、7.5m 高的巨大门洞，到达 2000m²、30m 高、顶部采光的共享中庭。为强调法律在市政生活中的权威性，面向城市的西立面以实体墙面上的凹洞为形式特征，南立面则设计为对称式。

The Beijing Higher People's Court consists of three main units: the trial building, the registration and petition building, and the security inspection hall. Other units include a 2,000 square meters hall, 15 court rooms, a law library, a 1,000 square meters space for legal service such as filing, registration and petition, as well as other ancillary facilities.

To realize the features of public space and the convenience of social services, the trial building is open to the outside ring road, which is set beside an open square next to the ring road. Visitors can enter this area from the square on the west side, stepping across a 70-meter wide, 4.5-meter high steps, going through a huge door with a 30-meter width and 7.5-meters tall, to reach the top atrium. Meanwhile, in order to emphasize the authority of law, the west facade, which faces the city, is characterized by a cavity in the solid wall, with south facade strict symmetry.

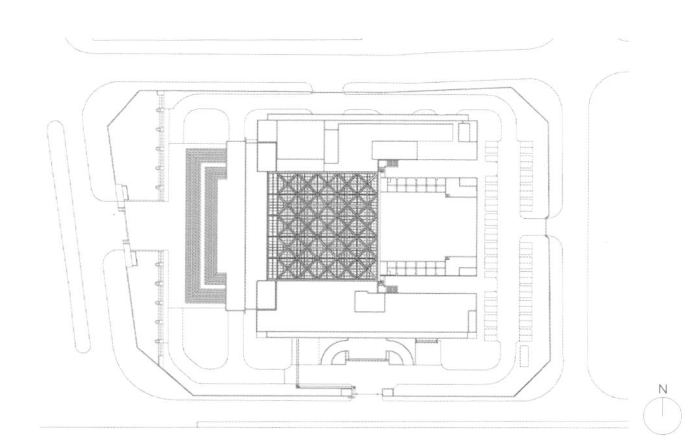

道路两侧

恒基中心
Henderson Center

建成时间 Completion Year：1997
建设面积 Building Area：300 000m^2
合作公司 Partner：香港关善明建筑师事务所
Simon Kwan & Associates

恒基中心工程被评为 1990 年代“北京十大建筑”之一，属于北京早期“混合使用中心”的建筑类型，功能多样，组织有序，协调互补。“中心”的内涵使其成为城市居民心目中的活动中心，而不是封闭的独立王国。其以北京火车站为中心，与站前街西侧已建成的建筑轮廓线相协调，与路北高耸的国际饭店相呼应，沿长安街的立面形象有一定标志性。

整体建筑既有现代化建筑的功能、体型，又有丰富、细腻的细节和经过抽象概括的古典片断，是一座“新古典”建筑。

The Henderson Center was honoured as one of Beijing's top 10 buildings in the 1990s. It was categorized as a “mixed-use center” building in early times, with various functions and orderly organization. The connotation of “center” makes it an activity center for the neighbourhhood residents, rather than a closed independent kingdom. The outlines of the Henderson Center echos to two building(s). To the north Zhangqian Street is the towering International Hotel. To the west of the street is the outlines of a group of buildings center around the Beijing Railway Station.

The Henderson Center has not only modernized functions and forms, but also many delicate details and pieces of abstract images reminiscent of the classic age, which can be called “new classical” architecture.

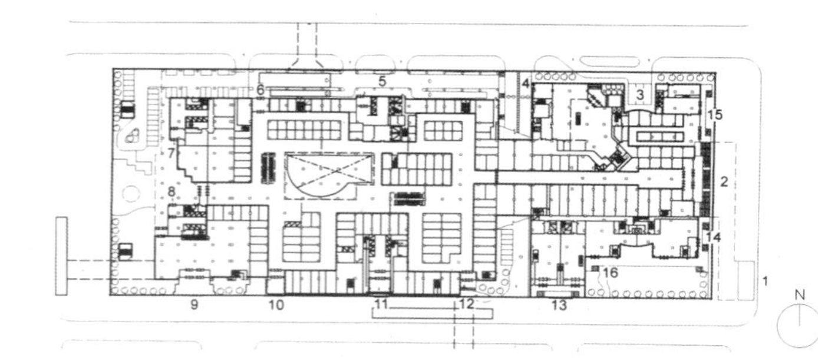

北京电视中心
Beijing TV Center

建成时间 Completion Year：2006
合作公司 Partner：日本日建设计 Nikken Sekkei / 中广电广播电影电视设计研究院 DRFT

北京电视中心将先进的电视工艺系统和整体建筑融为一体，是一座集楼宇控制自动化、办公自动化、信息传输自动化于一体的高科技含量的智能化大楼。主楼的中央部位设置了一个与主楼通高的大型中庭，平面围绕中庭形成开放式布局。外观上以中庭为中心，由东面呈阶梯状上升。主楼的超级大框架结构为国内第一座巨型钢架结构，坚固且有高度的灵活性。场地中的 3 栋建筑物通过设在基地中央的电视台广场而结为有机整体，形成协调的城市景观。

The Beijing TV Center integrates an advanced TV technology system into the entire project, which is a high-tech intelligent building with building-control automation, office automation, and information transmission automation. A large-scale atrium is set in the center, which is as high as the main building. And an open space surrounds the atrium. The large-scale frame structure of the main building is the first giant steel frame structure in China, which is strong and highly flexible. The three buildings are integrated as a unit through the TV Station Square in the center, forming an appropriate urban landscape at the same time.

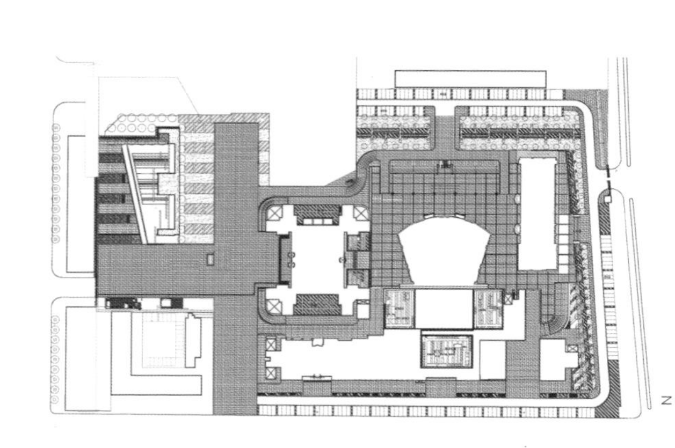

BTV

王府中环
Wangfu Central

建成时间 Completion Year：2017
建设面积 Building Area：149 633m^2
合作公司 Partner：KPF

项目地处北京最繁华的商业步行街——王府井大街，北侧紧邻北京市百货大楼，西侧遥望故宫，是北京市东城区政府制定的王府井商业发展规划中的重要一环，力图引进国际一线品牌专卖店和旗舰店，成为品牌和时装发布的标志性商业建筑，并助力整个王府井地区的改造与复兴。本项目包含超过 4 万m^{2}的零售空间和配有 73 间客房的文华东方酒店。由国际化的多个团队合作设计，采用国际通行的顾问制、建筑师负责制，实现设计—招标—施工全过程的控制。

This project is located along the most prosperous commercial pedestrian street — Wangfujing Street. Adjacent to the Beijing Department Store on the north, with clear views of the Forbidden City to the west, this project is the key to the Wangfujing commercial development plan formulated by the East District Government of Beijing. Thus, it strives to introduce international famous brand names and flagship stores, aiming to become an iconic commercial building for most up-to-date fashion mechandises, while at the same time making a contribution to transforming and revitalizing the Wangfujing block.

This project includes over 40,000 of retail spaces and 73 rooms for the Mandarin Oriental Hotel. Designed by an international team with multiple partners, which adopts an internationally accepted consultancy system and an architect responsibility system to realize the control of the whole design-tender-construction process.

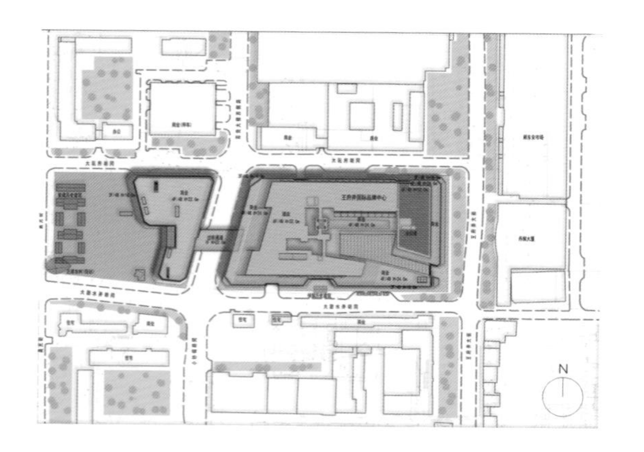

Marriott
VICTORIA'S SECRET
Superdry
VICTORIA'S SECRET

蓝色港湾
Blue Harbor

建成时间 Completion Year：2007
建设面积 Building Area：137 624m²
合作公司 Partner：JERDE

蓝色港湾包括配套商业用房、餐饮、文化休闲、娱乐、会展中心及配套服务用房等。项目在地下连成一体，地上由街道、下沉广场和半坡道路等划分为 19 个楼座，形成“村落”式布局。为了与周围环境相呼应，北面和西面的“都市围墙”试图表现北京的城市特质，由光亮的石板、混凝土及玻璃组成。靠近公园的区域采用了接近自然的表现手段、有机建材和景观。位于地下一层夹层中央的下沉花园将是阳光照耀的绿地，设有大型活动表演舞台及广场。

The Blue Harbor project includes commercial units, catering venues, cultural leisure facilities, entertainment venues, a convention center, and ancillary service rooms. Divided by street, sunken plaza and ramp-

ways, the 19 building blocks above ground form a village-like layout, with a connected underground space. In order to respond to the surrounding environment, a "Urban Wall" consisting of bright slate, concrete and glass in the north and west is set to express the city characteristics of Beijing. The area close to the park applies techniques that simulate as closely as possible the mother nature, along with organic building materials and landscaping. The sunken garden in the center of the mezzanine on the ground floor is a sunshiny green space with a large performance stage and a square.

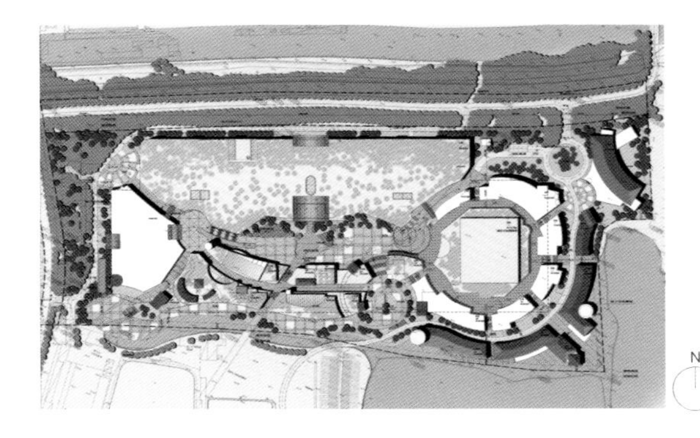

法国驻华大使馆新馆
New Embassy of France in China

建成时间 Completion Year：2011
建设面积 Building Area：19 950m²
合作公司 Partner：S.AREA ALAIN SARFATI ARCHITECTURE

法国驻华大使馆新馆位于朝阳区亮马桥路北的第三使馆区。新馆在体现法国设计特有元素的同时，将中、法文化兼收并蓄，充分融合在一起，形成内向型的现代四合院布局。建筑在形体与功能间进行了传统与现代的演绎，“城”与“楼”的组合树立并强化了建筑在区域中的控制力，以及在城市空间中的标识性。通过在室内外的细节处理中运用不同材质和色彩，比如简洁、精致的廊、檐、架的设计，创造出不同层次的灰空间。空间序列的排布结合了现代公共建筑和使馆建筑的特殊要求。

Located in the third embassy district to the north of Liangmaqiao Street, in Chaoyang District, the new embassy building of France in China reflects the unique and original element of French architecture while integrating the form of Chinese and French culture eclectically, forming a modern inward courtyard layout. The building interprets both the tradition and the modern in form and function. The combination of “city” and “building” establishes and strengthens the control in the area, as well as its unique identity in urban space. Assortment of materials and colors were applied in interior and exterior details, such as the simple, exquisite design of the corridors, eaves and frames, which creates various grey spaces. In the meanwhile, the arrangement of spatial sequences combines the special requirements of modern public buildings and embassy buildings.

三亚太阳湾柏悦酒店
Park Hyatt Sanya Sunny Bay Resort

建成时间 Completion Year：2015
建设面积 Building Area：66 706m^2
合作公司 Partner：DENNISTION INTERNATIONAL ARCHITECTS &PLANNERS LTD

太阳湾柏悦酒店由 6 栋主楼及裙房组成，整体布局依山就势，建筑高低错落，以争取最大的观海景观。建筑高度 46.10m，总客房数 218 套，合自然间 222 间。基于丰富多姿的自然环境，柏悦酒店以漂浮于海面之上的玉石作为概念原型。建筑布局化整为零，由数幢简洁的条状体量错落构成，分布于镜面水池之上。建筑立面的淡透乳白 U 玻在波光倒影的映衬下，营造出轻盈简洁的视觉形象。

酒店客房开间宽度为 6.6m，意在使客房的起居空间和卫生间均贴临外窗，进而享有开阔的海景。

Park Hyatt Sanya Sunny Bay Resort is composed of six main buildings and annex structures, with the overall layout responding to the natural surroundings, and the buildings of different heights offering grand views of the sea. The 46.10-meter tall building has 218 guest rooms and 222 natural rooms. Based on an ample natural environment, the design concept origins form the image of a piece of jade floating on the sea. The layout design divides the whole into parts, which is composed of several simple strip volumes distributed on the mirror pool. The facade, adopts a U-glass shape with a milky white that creates a light visual image under the reflection of sunlight.

The standard guest room is 6.6-meters wide, which is designed to make the living space and bathroom close to the window in order for guests to enjoy grander sea views.

昆仑饭店
The Kunlun Hotel

建成时间 Completion Year：1986
建设面积 Building Area：80 000m²

昆仑饭店是 1980 年代初由国内设计、用于接待外国旅客的规模最大的旅馆之一。主楼平面结合用地呈 S 形，北边凹处设旅馆门厅，南边凹处设四季厅餐厅、休息廊等，与南边的亮马河岸和绿地呼应。主楼最高点 102m，西翼 21 层，东翼 24 层，中央塔楼 28 层。屋顶设有圆形旋转餐厅和直升飞机停机坪。客房有 1005 间单间，7 组三间套房和 1 组中国宫廷式豪华套房。公共部分设在 1~2 层，一条曲折的小街将餐厅、酒吧、茶座、游泳馆和咖啡厅等空间串联起来。

Designed by domestic architects in the early 1980s, the Kunlun Hotel was one of the largest hotels in Beijing aiming to receive foreign tourists. The layout plan of the main building is S-shaped with a lobby in the north, a restaurant and leisure corridors in the south, responding to the Liangma riverbank and green land. The highest point of the main building is 102 meters, with 21 floors in the west wing, 24 floors in the east wing and 28 floors in the central tower. A circular revolving restaurant and helicopter apron are set on the roof. There are 1,005 guest rooms, 7 suites, and 1 deluxe suite with a traditional Chinese palace style. The first two floors are public spaces and host a winding indoor street that connects restaurants, bars, tearooms, a swimming pool and cafes.

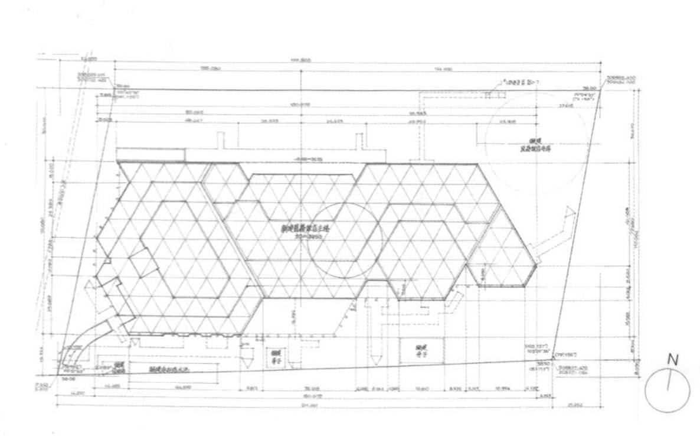

KUNLUN HO

钓鱼台国宾馆会议中心
Conference Center of Diaoyutai State Guesthouse

建成时间 Completion Year：2013
建设面积 Building Area：22 800m^2

设计用地位于钓鱼台国宾馆东南角，主入口设在北侧，并留有足够的广场空间。门厅、序厅、宴会厅构成南北序列，贵宾区、过厅构成东西序列。大宴会厅和大报告厅设置在建筑南部，最大程度地将自然美景引入室内。序厅与门厅结合，将人们的视线引向北侧的丹若园。二层的核心是南部的大报告厅，其西侧是长长的玻璃序厅，可远观美丽湖景。项目从古建区汲取灵感，采用暖黄色石材、深灰屋顶和暗红木纹转印框材等。同时，舒缓的双重檐坡屋顶也降低了建筑的尺度感。这些都使会议中心与园区内的古建筑达到和谐与统一。

The conference center is located in the southeast corner of the Diaoyutai State Guesthouse. The main entrance is set on the north side of the compound with a spacious square. The entrance hall, the preface hall and the banquet hall constitute a space sequence from north to south, while the VIP area and the secondary hall constitute the space sequence from east to west. The banquet hall and the lecture hall are set in the south part to provide maximized natural views. The preface hall is combined with the entrance hall, which attracts people's views to the Danruo Garden. The core space of the second floor is a lecture hall with extensive glass volume where people can enjoy a beautiful lake view. The project draws inspiration from traditional architecture, so yellow stones, a dark grey roof and dark red wooden frame are adopted. In the meanwhile, the double-sloping roof makes the space scale more humane. The design brings a sense of harmony and unity to the conference center and surrounding ancient buildings in the park.

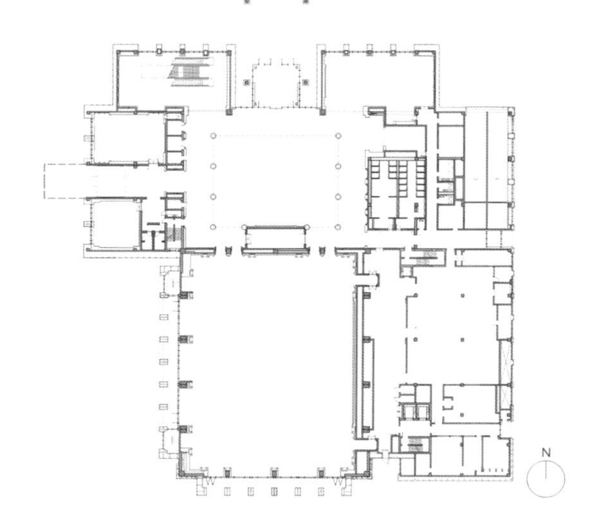

昆明长水国际机场航站楼
Kunming Changshui International Airport Terminal Building

建成时间 Completion Year：2012
建设面积 Building Area：548 440m^2
合作公司 Partner：Arup

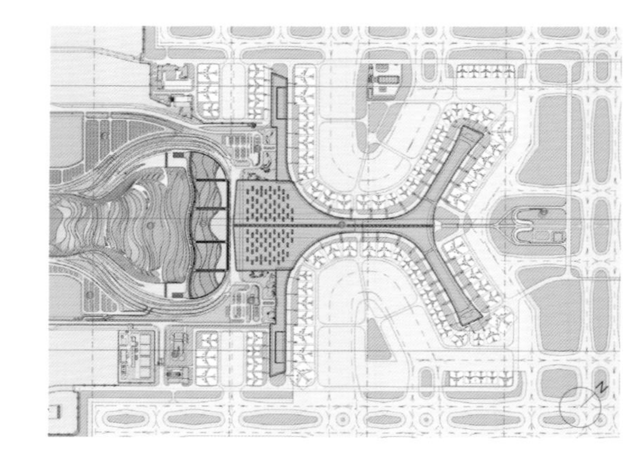

昆明长水国际机场航站楼可达到 2020 年旅客吞吐量 3800 万人次的预测目标。按照国内国际分开，出发到达分流的原则，设计以便捷的流线、完善的布局体现了现代大型枢纽机场的功能性需求。航站楼总高度 72.9m，在平面构型上可分成前端主楼中心区和 5 条候机指廊。主楼为地上 3 层（局部 4 层）、地下 3 层构型，利用航站区南低北高的自然地势，设计为空陆侧错层，减少航站区土方工程。航站楼采用中心区减隔震、自然通风、自然采光等，是国内首个获得三星级绿色建筑设计标识证书的机场航站楼项目。

同时，建筑也充分体现了多彩云南的地域特色，翘曲的双坡屋顶表现了当地民族传统建筑的神韵，构成了航站楼最显著的特色。室内支撑结构采用了钢结构“彩带”，7 条彩带沿南北方向有序展开，不仅将航站楼离港层主要功能区予以划分，也以卓越的科技力量将现代美学与地域文化特色完美统一，成为整个建筑立面和内部空间的标志。

The Kunming Changshui International Airport terminal can accommodate the target of 38 million passengers throughout 2020. According to the principle of the separation of domestic and international airlines, and the seperation of departure and arrival, the project meets the functional requirements of a modern large-scale airport hub with convenient procession and a reasonable layout. The height of the terminal is 72.9 meters, which can be divided into a main passenger area and five finger corridors. The main building has three above-ground floors (four floors in some areas) and three underground floors. Taking advantage of the natural terrain which is lower in the south and higher in the north, the design separates the airside and landside floors into different levels, so as to reduce the earthwork. The terminal was honored the pioneer project in China to obtain the “three-star green building design label”, which adopts environmental protection measures, such as central area seismic isolation, natural ventilation, and natural lighting.

At the same time, the building also embodies the regional characteristics of colorful Yunnan. The warped double-slope roof responds to the charming traditional architecture of China, which has become the most significant feature of the terminal. The interior structure adopts steel structure ribbons, with 7 ribbons set along with the north-south order, which not only divides the main functional areas of the departure floor but also unifies the modern aesthetics and regional culture with outstanding technology, highlighting the symbol of the facade and internal space.

3
3
2

广州南站
Guangzhou South Railway Station

建成时间 Completion Year：2010
建设面积 Building Area：590 396m²
合作公司 Partner：中铁第四勘察设计院 China Railway Siyuan Survey and Design Group/TFP

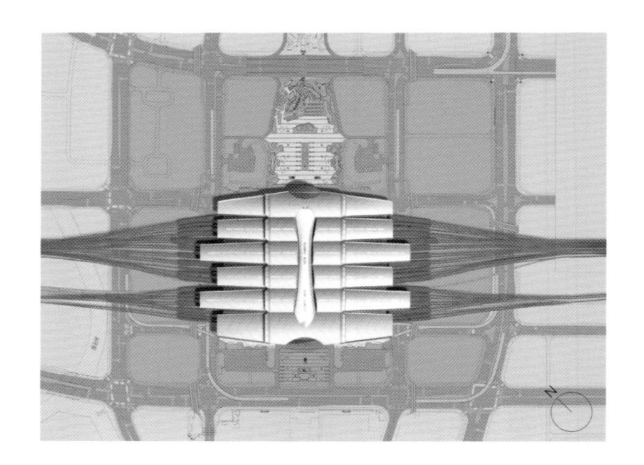

广州南站是全国四大客运枢纽之一，选址于广州市番禺区，车站设计旅客发送量近期为7670万人次/年，远期为11 075万人次/年。新广州站总面积达48万 m^2，包含15条铁路站台、28条到发线，是国内第一个高架铁路车站，集客运专线、城际铁路、普通铁路、城市轨道交通、海关、机场客运、巴士、出租、社会车等多种交通方式于一体，实现了真正意义上的交通零换乘。

车站造型新颖，屋顶以层叠的蕉叶状的单元为造型的主体，体现了岭南的地域特征。车站共有五层，地上三层，地下二层。即地面出站层、地面站台层、高架候车层、地铁站台层、地下停车层，此外高架层上预留商业夹层，地下停车层中间为地铁站厅。旅客流线采用上进下出和下进下出相结合的方式。

Located in Panyu District, Guangzhou South Railway Station is one of the four major transport hubs in China. The station is designed with the passenger volume of 76.7 million people per year in the near term and 11.75 million people per year in the long term. The overall cross-area of the New Guangzhou Station is 480,000 square meters, and includes 15 railway platforms and 28 arrival-departure lines. It is the first station in China with elevated railways. The building integrated the passenger transportation lines, intercity railways, ordinary railways, urban rail transit, customs, airport shuttle lines, long-distance bus, taxi, and other transportation means, realizing an ideal connection without transfer.

The unique and novel architectural form with a laminated leaf-like shape, reflects the regional characteristics of the Lingnan area. The building has five floors with three above-ground floors and two underground floors, including arrival platforms, departure platforms, an elevated waiting space, metro platforms, and underground parking. In the meanwhile, there is a mezzanine layer for commercial use on the elevated floor. The metro station hall is set in the center of the underground parking space.

首都医科大学附属北京天坛医院
Beijing Tiantan Hospital, Capital Medical University

建成时间 Completion Year：2018
建设面积 Building Area：352 294m²
合作公司 Partner：Beeg Geiselbrecht Lemke Architekten Gmbh

天坛医院新院区由一组先进的生态医疗建筑群组成，整体布局形似超级航母，象征承载着全国神经外科领域的研究与发展。整体造型动势极强，如欲乘风破浪、扬帆远航。

总体布局体现由外向内、由动到静的特点，在切合医疗流线的同时，为病患提供更加优质与完善的使用空间。设计团队按照功能设置多个出入口，便于集散，病患可直接由室外进入目标科室，简化诊疗流程，也可避免人员拥挤，降低交叉感染的几率。医疗功能紧密相连的科室就近布局，缩短了医患路线。医疗区在南北向设置高效、宽敞的绿色通道，门诊、急诊、

医技、病房区域通过这条通道来组织区域内部交通。东西方向对应着三栋病房楼的位置，设计了三条快速通道，使门诊、急诊、医技和病房区域之间的交通更加便捷。

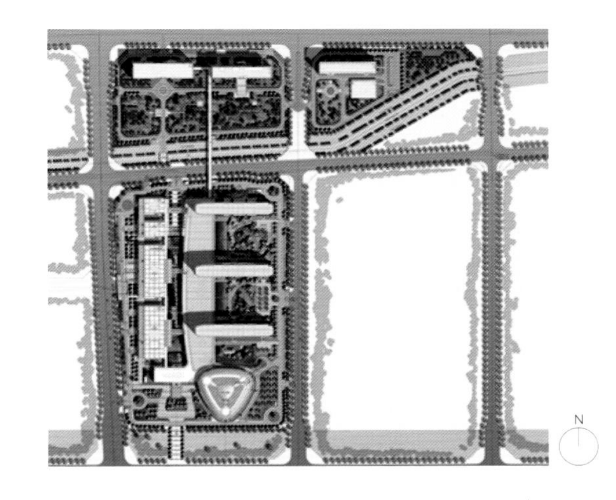

The new Tiantan Hospital building cluster is composed of a group of advanced ecological medical buildings. The overall layout resembles a super aircraft carrier, symbolizing its enormous responsibility for the research and development in the studies of neurosurgery.

The overall layout reflects the characteristics from outside to inside, from dynamic to static. On balance, the arrangement of circulations for various functions of an hospital has taken full consideration of providing patients with high-quality space. The architects set multiple entrances and exits according to different functional demands. Patients can enter his ward or lab without much trouble to simplify the treatment process. The design can also avoid crowding and reduce the probability of cross-infection. The medical departments are closely linked in order to shorten the distance for both doctors and patients. The medical area is equipped with an efficient green passageway from north to south to organize intra-regional traffic for the outpatient, emergency, medical, and ward areas. Three east-west passageways that connect the three ward buildings help solve the problem of complicated circulations between the clinic, emergency, medical and ward area.

深圳宝安国际机场 T3 航站楼

Shenzhen Baoan International Airport Terminal 3

建成时间 Completion Year：2013
建设面积 Building Area：451 000m^2
合作公司 Partner：Massimiliano & Doriana Fuksas

深圳宝安国际机场 T3 航站楼地处滨海。建筑试图与自然环境融合，如同生物一样成为自然界的组成部分。建筑形象的灵感来自海洋生物蝠鲼从水中跃起的瞬间。建筑由双层表皮系统包裹，金属板与玻璃间隔布置，白色的金属材质与花纹玻璃相间形成均匀的纹理，建筑的体量如同由半透明的白纱包裹。T3 航站楼主要由主楼中心区、中央指廊区、指廊中心区、东翼指廊区、西翼指廊区、十字北指廊、十字东指廊及十字西指廊组成。主楼大公共空间采用大跨度柱网，指廊区出发层为数百米长的连贯无柱空间。

Located at the seaside, Shenzhen Baoan International Airport Terminal 3 attempts to integrate the natural environment to become a part of the natural world. The inspiration of the architectural form originates from the image of the moment when the manta leaps out of the sea. The building is wrapped by a double-layered skin system, with metal plate and glass arranged at intervals. White metal material and figured glass constitute a uniform pattern. The image seems to be wrapped in a translucent tulle. The building hosts the center area, finger corridor area and its central space, finger corridors on the east side and the west side, and crossed corridors on the north, east and west sides. The large-scale public space of the main building adopts a large-span column network, and the departure floor is a continuous column-free space which is hundreds of meters long.

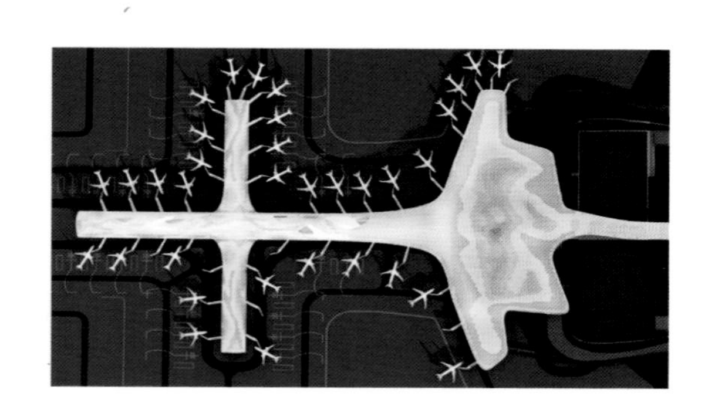

349
350
350A

青岛北站

Qingdao North Railway Station

建成时间 Completion Year：2014
建设面积 Building Area：68 828m^2
合作公司 Partner：中铁二院工程集团有限责任公司 China Railway Eryuan Engineering Group / AREP

青岛北站是集铁路客运、长途汽车、城市轨道交通及常规公共交通于一体的山东省最大的立体综合交通枢纽。站场设计为8场16线，最多聚集人数1万人。在功能布局上系统地考虑了铁路站场与站房、站前广场、公交站、长途车站、地铁等之间的交通和功能关系，提出“以流为主、到发分离、南北贯通”和“无缝对接、零换乘”等理念，并采用立体分流的模式提高换乘效率。以简洁有力的结构语言隐喻展翅的海鸥。

高架候车大厅内开敞通透，10榀钢斜拱模拟海鸥翅膀扇动时的不同位置，上覆轻型金属屋面，形成动态的建筑造型。

The Qingdao North Railway Station is the largest comprehensive transportation hub in Shandong province, with integrated railway transport, long-distance buses, urban transit and other conventional public transportation. With eight platforms and sixteen bus lines, the building can accommodate about 10,000 people at a time. The design considers the functional connections between depots, platforms, the open square in front of the station, local bus stations, long distance bus stations, and subway stations. The key principles of the overall design include the separation of departure and arrival circulations, connecting spaces from north to south, to create a seamless transition environment. The structural language of the building is simple and powerful: to present itself an outline of a flying seagull.

The waiting hall is an open space with 10 steel oblique arches, simulating different positions when the seagull waving its wings, and it is covered with a light metal roof, making a dynamic architectural form.

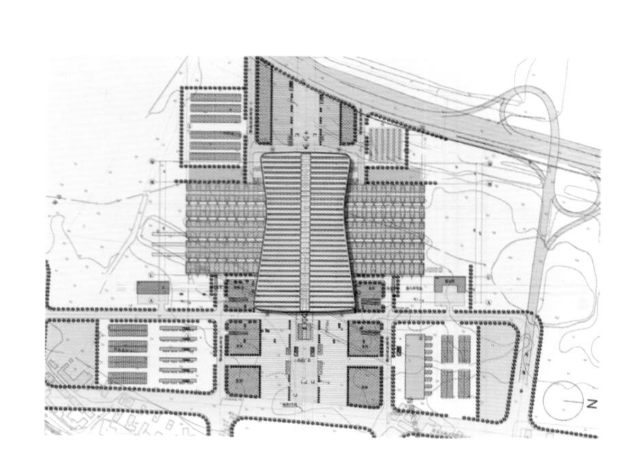

开水
卫生间
14
A6·A7
卫生间
B6 B7
检票口 B8-B1
B4·B5 检票口

南京南站主站房

Main Station of the Nanjing South Railway Station

建成时间 Completion Year：2011
建设面积 Building Area：281 021m^2
合作公司 Partner：中铁第四勘察设计院
China Railway Siyuan Survey and Design Group

南京南站是集铁路客运、长途汽车、城市轨道交通及其他城市常规公共交通于一体的大型综合交通枢纽。

站房建筑整体划分为竖向交通、卫生间、多功能集成模块、外幕墙、内幕墙、清水混凝土构件、地面铺装、楼面及屋面吊顶、屋面、贵宾室、静态标识等若干系统。各部分相对独立，使系统化的设计分工得以实现。

将中国传统建筑组群注重中轴序列的特点应用到候车大厅的设计之中。立面将传统木构造型与现代结构技术巧妙结合，创造出承力斗拱，形成了富有新意的檐下空间。

The Nanjing South Railway Station is a large-scale transportation hub which consists of railway, bus, and urban rail transit systems.

The functions of the Station include various sub-systems such as vertical traffic, sanitary, external and interior curtain walls, pure concrete components, ground pavement, floors and ceilings, VIP rooms and a guidance system. Sub-systems are relatively independent to make possible the systematic division in design.

The design also applies the feature of emphasizing the axial sequence of a traditional Chinese architecture to the design of the station's waiting hall. The traditional wooden structure of the facade is built with modern techniques, a load-bearing *Dougong*, to form a unique space under the eaves.

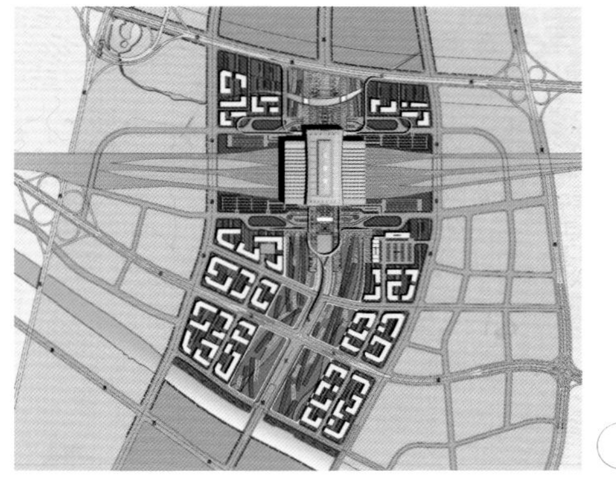

北京儿童医院门诊楼
Beijing Children' s Hospital Outpatient Building

建成时间 Completion Year：2003
建设面积 Building Area：37 000m²

儿童医院门诊楼位于医院正门的东南侧，建筑面积 15 000m²，包括儿科急救中心及外宾儿科医疗、儿科研究所、保健所。急救中心平面为矩形，地下 2 层，地上 9 层，研究所、保健所平面呈“凹”字形，地下 2 层，地上 5 层。两楼间以连楼相接，内部采用双走道、多跨度、小天井的布局，功能分区合理。外立面为孔雀蓝琉璃板屋檐和白色面砖墙面，简洁淡雅。主体结构为现浇钢筋混凝土框架剪力墙体系，楼内设有空调、自动消防喷洒以及闭路电视等设施。

The Beijing Children's Hospital Outpatient Clinic is located to the southeast of the main entrance with a total gross floor area of 15,000 square meters. The building is home to the pediatric emergency center, the pediatric medical center for foreign guests, the Pediatric Research Institute, and the health care institute. The layout of the pediatric emergency center is in rectangular shape with nine above-ground floors and two underground floors. The layout of both research institute and health care institute were designed in a concave shape, with five above-ground floors and two underground floors. The two buildings are tightly connected. Double walkways, multi-span spaces and small courtyards have been implemented to create reasonable function distribution. The facade is made of white brick walls with peacock blue glazed eaves, which is simple and elegant. The structure system consists of steel-reinforced concrete frames with shear walls. The building is also equipped with air conditioning, automatic fire sprinkler, closed-circuit television and other facilities.

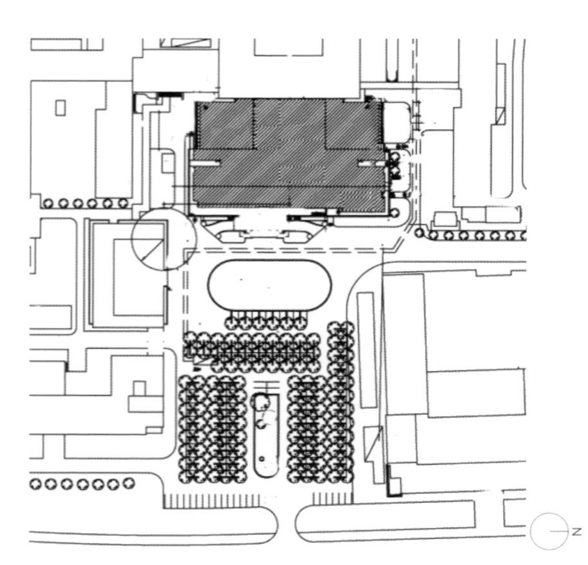

北京儿童医院
Beijing Children's Hospital
门诊
Outpatient Department

首都医科大学宣武医院改扩建一期工程

Xuanwu Hospital Capital Medical University Reconstruction and Extension Phase I

建成时间 Completion Year：2018
建设面积 Building Area：80 675m^2
合作公司 Partner：hks

本项目地上建筑包括中国国际神经科学研究所、干部保健楼和附属用房三栋建筑。研究所以德国国际神经科学研究所为母版，建筑形态象征人体大脑，布置于城市道路转角 2.4m 高的基座上，四周是斜坡式绿化庭院，为地下空间带来采光通风的便利条件。附属用房位于研究所西侧，采用简洁现代的正方形庭院式布局。干部保健病房楼布置在研究所和附属用房的北侧，南立面各楼层用舒展的弧线加以突出，产生刚柔相济的效果，与研究所的椭圆造型形成呼应。

The above-ground buildings of the project host the China International Neuroscience Research Institute, Cadre's Health Building and subsidiary building. Taking the German International Institute of Neuroscience as master reference, the architectural form was designed as a human brain, placed on a 2.4-meter base at the turning corner along the street. It is surrounded by a sloping green courtyard, which brings lights and ventilation to the underground space. The subsidiary building, located on the west side of the institute, features a simple, modern and square-courtyard style. The Cadre Health Building is located on the north side of the research institute and the subsidiary building. All the floors from the south elevation are highlighted with extending curves, which create an effect of harmony between hardness and softness, as well as naturally transiting into the elliptical shape of the institute.

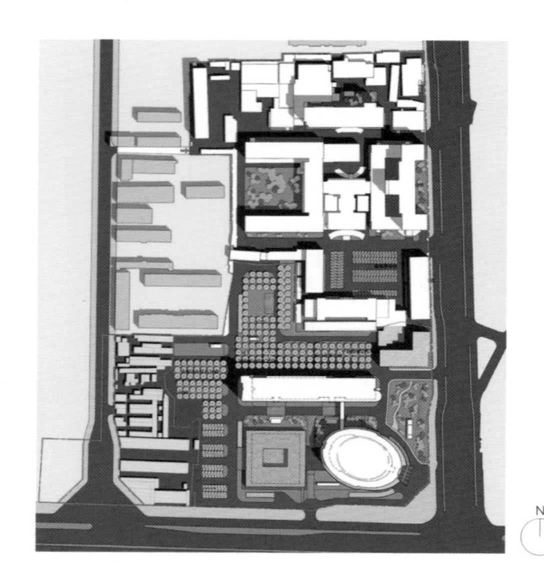

首都医科大学宣武医院
Xuanwu Hosp al Capital Medical Univers y

居住建筑
RESIDENTIAL BUILDING

226 万科第五园

The Fifth Vanke Village

230 北京 SOHO 现代城

SOHO New Town Beijing

232 万科中粮假日风景（万恒家园二期）
D 地块 D1、D8 工业化住宅

Parcel D1 and D8 Industrialized Housing of Vanke
COFCO Holiday Town (Wanheng Homeworld Phase Ⅱ)

234 建国门外外交公寓

Jianwai Diplomatic Residence Compound

万科第五园

The Fifth Vanke Village

建成时间 Completion Year：2005
建设面积 Building Area：125 000m²
合作公司 Partner：澳大利亚泊涛建筑设计公司
PT Architecture Design

万科第五园位于深圳市龙岗区布吉镇坂雪岗南区，力求使用现代手法对中式传统住宅进行演绎。整个社区由中央景观带分隔成两个“村落”，每个“村落”都由三种产品——庭院House、叠院House以及合院阳房所构成。各“村”内部都由深幽的街巷或步行小路以及大小不同的院落组合而成。紧邻城市干道的商业街和社区图书馆与住宅区以池塘相隔，通过小桥相连，互为景观，其内部空间也特别强调各种开敞、半开敞、下沉的院落和连廊的组合，形成丰富而使人流连的“村口”场所。

庭院别墅的“前庭后院中天井”以及通过组合形成的“六合院”和“四合院”，叠院House的“立体”小院（院落＋露台），合院阳房围合成的“大院”，种种院落形式体现了传统民居的“内向”型空间。该项目吸收了广东特色的竹筒屋和冷巷的传统做法，通过天井、廊架、挑檐、高墙、花窗、孔洞、缝隙、窄巷等，房屋在梳理阳光的同时呼吸微风，住宅舒适度提高的同时有效降低了能耗。

The Fifth Vanke Village is located in the south of Banxuegang, in Buji Town, Longgang District of Shenzhen. The design strives to take the modern approach to interpret traditional Chinese residential dwelling. The entire community consists of two "villages" separated by a central landscape belt. Each "village" is composed of three typologies — the Garden House, Stack House and Courtyard House. Inside each "village" is a cluster of extensive streets, quiet pedestrian paths and yards on different scales. The commercial streets and community library that sit close to the urban trunk roads are separated from the residential areas by ponds and are connected by small bridges, complementing each other as part of a landscape. The interior space also emphasizes the combinations of open, semi-open, sinking courtyards and corridors, forming a rich and welcoming "village entrance."

Different forms of courtyards, from the "lobby-courtyard-garden" in courtyard villa, courtyard with four units and with six units, to the "three-dimensional" stacked yards (yard and terrace), the "big yard" enclosed by sun-faced units, reflecting the inward character in traditional residential spaces. The design adopts the traditional way of building bamboo-tube houses and "cold lane" in Guangdong, through patios, porches, overhanging eaves, high walls, lattice windows, openings, cracks, narrow alleys, and many other concepts, that bring in the breeze and the sunlight, reducing energy consumption while improving the living experience.

北京 SOHO 现代城
SOHO New Town Beijing

建成时间 Completion Year：2001
建设面积 Building Area：233 025m²

北京 SOHO 现代城是集办公、公寓、购物、娱乐于一体的大规模城市建筑群。建筑东西向长达 280m，主体高度均在 60m 以上。在考虑经济效益和使用要求的基础上，将主体由东到西划分为 A、B、C、D 4 座单体建筑。东端 A 座是 132m 的超高层建筑。4 座建筑均以正方形平面和立面构图，合理、经济、实用。同时，在 4 个建筑之间留出宽 25m的 3 个巨形空洞，阳光和暖风穿过建筑群，缓解了大型建筑与城市街道之间的矛盾，避免了“街墙”效果。

将 A 座避难层扩大，并和 D 座设备层以及 B 座、C 座屋顶连接在一起，以流动的曲面屋顶形成一个 3 层高度的绿色长廊。

SOHO New Town Beijing is a large-scale urban architectural complex integrating offices, apartments, shopping and entertainment. The east-west section of the building is 280 meters long and the general height of the main part is above 60 meters. Considering the economic benefits and functional requirements, the main volume of the complex is divided into four isolated buildings, which are Block A, Block B, Block C and Block D, which run from east to west. Block A at the east end is a high-rise building with a height of 132 meters. The four buildings are all in square volume, which has a reasonable, economical and practical effect. At the same time, there are three giant hollows with a width of 25 meters between the four buildings. The sunlight and warm winds pass through the buildings, alleviating the contradiction between large buildings and streets and avoiding the effect of “street walls”. We expanded a refuge floor in Block A, connected it to the equipment floor of Block D and the rooftop of Block C, creating a 3-story green corridor with a streamline curving roof.

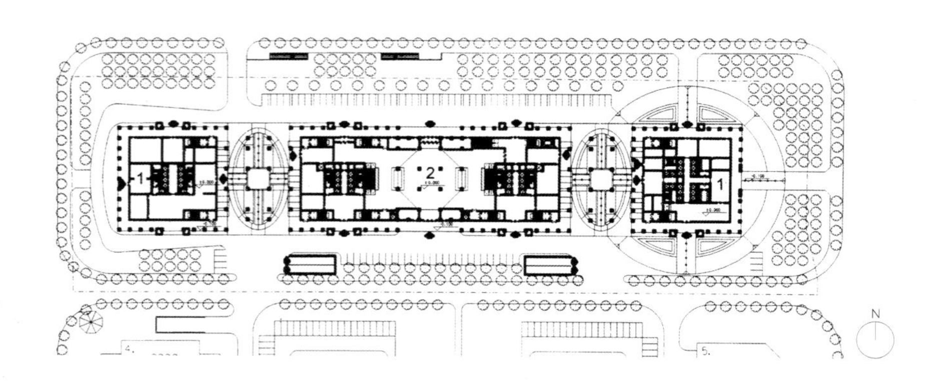

SOHO现代

万科中粮假日风景（万恒家园二期）D 地块 D1、D8 工业化住宅

Parcel D1 and D8 Industrialized Housing of Vanke COFCO Holiday Town (Wanheng Homeworld Phase Ⅱ)

建成时间 Completion Year：2010
建设面积 Building Area：33 000m²

该项目采用装配整体式剪力墙结构体系，改进和完善了保温装饰承重一体化外墙。通过对预制外墙构件进行拆分、组合，将立面划分为预制外墙、阳台挂板和金属部品三个部分。预制外墙是结构的一部分，灵活度较差，而形成整体的凹凸关系，是立面形式的基础。阳台挂板和金属部品配置的灵活性较高，因此用它们配合整体的需要。南立面全部用预制装配的方式构成，形式上凸显“浇筑”和“塑造”的特征。外立面表现形式依托工业化预制体系，充分体现出混凝土在住宅建筑中的表现力及可塑性，体现出“塑造、装配、清水”三特征。

This project adopted an integral shear wall structure system to improve and perfect the integrated exterior wall for insulation, decoration and loadbearing. By splitting and combining the prefabricated exterior wall components, we divided the facade into three parts: prefabricated exterior wall, hanging plates of the balcony and metal parts. The prefabricated exterior wall, as part of the entire structure, is comparatively low in flexibility and has formed the overall concave-convex elevation; the balcony panels and metal components are highly flexible. The south facades are all constructed in a prefabricated manner, manifesting the features of “casting” and “shaping”. The expression of the facade relies on the industrial prefabrication system, which fully reflects the expressiveness and plasticity of concrete in residential buildings, bringing out the three characteristics of “sculpturing, assembling and pure concrete”.

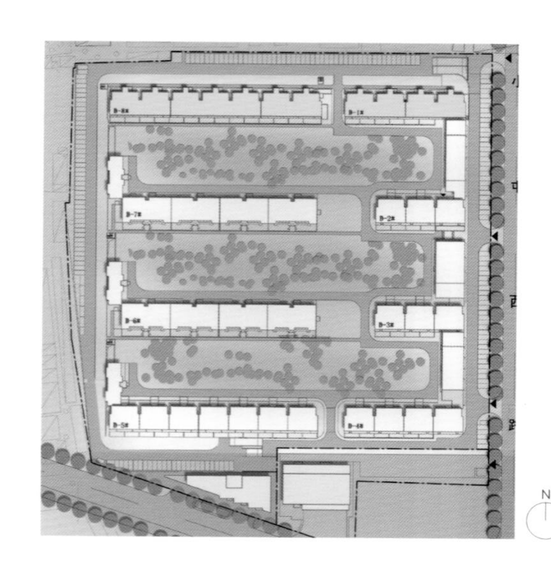

建国门外外交公寓
Jianwai Diplomatic Residence Compound

建成时间 Completion Year：1984
建设面积 Building Area：164 000m²

建国门外外交公寓位于建国门立交桥东北侧长安街与二环路交汇处，紧邻 CBD 商圈，地理位置优越，交通便利。项目始建于 1971 年，隶属于北京外交人员房屋服务公司，为各国驻华使馆、国际组织代表机构、各新闻机构及其人员提供办公、住宅用房和相关的服务。

Jianwai Diplomatic Residence Compound is located at the intersection of Chang'an Avenue and Second Ring Road on the northeast side of Jian Guo Men Overpass, close to CBD area, with superior geographical position and convenient transportation. Built in 1971, the project is affiliated with Beijing Personnel Service Corporation for Diplomatic Missions, which provides various categories of permanent and temporary service personnel for the Diplomatic Missions, Representative Offices of International Organizations and offices of foreign news agencies in Beijing.

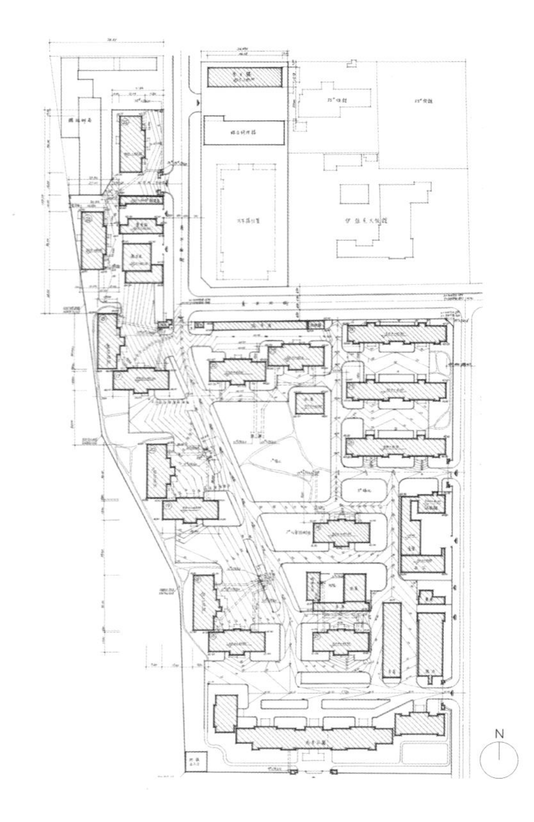

城市设计及其他
URBAN DESIGN AND OTHERS

238 丽泽金融商务区城市设计
Urban Design of Lize Financial Business District

242 怀柔科学城城市设计
Huairou Science City

246 西安唐大明宫国家大遗址保护展示示范园区
Xi'an Daming Palace National Heritage Protection and Demonstration Area

250 前门东三里河复兴
Qianmen Dongsanli River Regeneration

丽泽金融商务区城市设计

Urban Design of Lize Financial Business District

设计时间 Design Began：2009
规划面积 Planning Area：2.81km^2
合作单位 Partner：SOM 建筑设计事务所 Skidmore, Owings & Merrill LLP / 波士顿咨询（上海）有限公司 The Boston Consulting Group

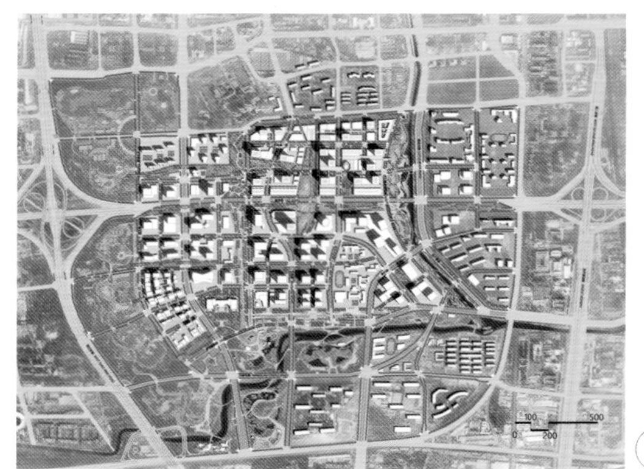

项目位于北京市西南二环路和三环路之间。规划方案紧密围绕新总规的落实，着力打造人本城区、紧凑城区、绿色城区、活力城区，将丽泽金融商务区建设成为支撑首都现代服务业发展的重要功能区和现代化大都市高品质建设的典范区。

明确功能定位。促进京津冀协同发展，丽泽航站楼与新机场实现轨道直连，实现五条轨道线的便捷换乘；补齐公共服务设施短板。

加强轨道交通车站地区功能、交通、环境一体化规划；优化轨道线网结构，服务城市通勤出行；建设“小街区，密路网”的开放街区，提升步行体验；坚持公交优先；加强职住平衡。

保障首都政治安全，落实中心城区高度管控要求，区域总体高度控制在 200m 以下；打造具有历史文脉和人文关怀的新金融中心；建设绿色城区，突出生态优势，完善 34 公里的生态慢行系统等。

This project is located between the Second Ring Road and the Third Ring Road in southwest Beijing. The scheme closely focuses on the implementation of new master planning and strives to build a humanism-based city, a compact, green and dynamic city, and to establish the Lize Financial Business District as part of an important model area that will support the development of Beijing's modern service industry and the establishment of a high-quality modern metropolis.

Clear functional positioning. To promote the coordinated development of Beijing-Tianjin-Hebei, the existing terminal of Lize Airport is connected to its new extension by rail transit so passengers can transfer between five rails with high convenience, which greatly improved the overall performance of public transportation.

We must strengthen the integrated planning of functions, transportation and environment in the area of rail transit station; optimize the rail network structure to facilitate daily commutting; build an open district of "small streets and dense road networks" to enhance the walking experience; prioritize public transportation; strengthen the jobs-Housing balance.

To guarantee the political security of the capital, we must also implement building height controls in central districts, where the overall height of buildings should be lower than 200m, establish a new financial center with historical context and humanistic qualities, build a green city to highlight ecological advantages, and complete the 34-kilometer ecological slow-traffic system.

怀柔科学城城市设计
Huairou Science City

设计时间 Design Began：2019
规划面积 Planning Area：100.9km^2

怀柔科学城是落实全国科技创新中心建设的重要战略性项目，项目围绕科学、科学家、科学城三个核心要素的内在关系，思考城市如何为科学服务、为科学家服务，成为匹配国家战略的世界级原始创新承载区，思考城市以怎样一种新形态与自然山水相融共生，树立生态涵养区、浅山向平原过渡区的城市营建新标杆。

着重从生态、功能、交通、风貌四个层面提出核心策略：蓝绿交织孕育科学灵感——将生态山水通过活力绿脉以及不同层级服务半径的开放空间引入城市，建设鼓励交往的公共空间环境。复合功能激发城市活力——通过规划结构引导、用地功能混合、楼座功能混合等方式来最大化地混合城市功能。人本网络构建宜人空间——着眼人本尺度，建设尺度宜人的小街区、密路网。山城融合塑造特色风貌——突出浅山地区自然景观特征，建立山城交融、起伏有韵的城市天际线；塑造城绿交融、疏密有序的城市空间形态。

Huairou Science City is an important strategic project for the establishment of the National Science and Technology Innovation Center. Focusing on the internal relationship between the three core elements of science, scientists and a science city, we tried to figure out how the city can serve science, scientists and the nation's strategic vision, how the city can develop together along with nature. We tried to establish a new benchmark for city buildings, where ecological conservation areas and transition areas between shallow mountains and plains are set up.

We put an emphasis on ecology, function, transportation and culture, and we proposed our core strategy:

Intertwining blue and green to inspire science: introducing the ecological landscape through dynamic green belts and open spaces in different layers and service radius.

Stimulating the vitality of the city by composite functions: we need to maximize the function of the mixed city by planning structure guidance, mixing of land uses and mixing of building functions.

We must construct friendly spaces with a human-oriented network to focus on the human scale and to build small blocks and dense networks of roads.

Building a distinctive culture through combining city and the mountains to establish a fluctuating skyline of intertwined city and mountains while highlighting the natural landscape features; to create spatial forms where the landscape and the city are connected properly.

西安唐大明宫国家大遗址保护展示示范园区

Xi'an Daming Palace National Heritage Protection and Demonstration Area

建成时间 Completion Year：2010
规划面积 Planning Area：3.7km^2

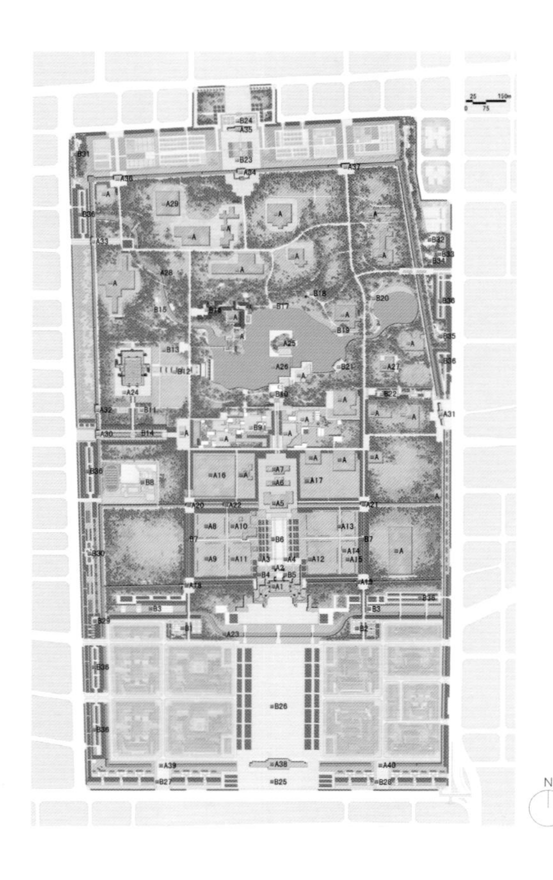

项目分为殿前区、宫殿区、宫苑区（包括太液池）及缓冲区。总体设计以突出遗址为战略定位，功能、空间、景观的打造均围绕文物保护主题，是关联文物保护、规划、建筑、景观及市政道路的综合且复杂的系统工程。因此，遵照中国和国际文化遗产保护的相关准则，在遗址公园内进行考古、保护、展示及建设。总体景观针对其地表仅存少量夯土遗迹、大部分遗址埋藏于地下的特点，不做复原性建设。通过将地面遗迹保护性展示、地下遗址标示性展示、浅表层的可逆性游览系统与确无遗址区域的绿化相结合，引导城市绿茵功能的适量融入。

The total area of the project is 3.7 km^2, which is divided into the pre-palace zone, the palace zone, palatial garden (including the Taiye pool) and the buffer zone. In our strategy, the overall design highlighted the heritage, functions, spaces and landscapes, which were all built around the theme of cultural relics protection, making it a complex project which involves cultural relics protection, urban planning, architecture, landscape and municipal roads planning. In accordance with relevant guidelines on the protection of Chinese and international cultural heritage, archaeological digging, protection, display and construction were carried out in the park. Considering the fact that most of the relics are buried, little terracotta remains, we cancelled restorative landscape construction, and instead introduced modest greenery into the site to cope with the protective display of aboveground relics, an identifiable display of underground relics and a reversible tour system on a shallow layer of earth.

前门东三里河复兴
Qianmen Dongsanli River Regeneration

建成时间 Completion Year：2016
规划面积 Planing Area：180 000m²

项目为距离天安门最近的历史文化保护区之一。运用先进的规划理念，结合城市、围绕河流，合理开展整体概念设计。以“老胡同，新生活”为理念，从“文化复兴”目标出发，引入“人居环境”科学及“城市复兴”理论，提出城市减灾、民生改善、风貌重塑、城市织补、空间提升、文脉传承、生态修复、设施完善八大规划理念。保留原有胡同规制和道路肌理，改善生活环境，进而提高居民生活品质。修复前门三里河水系，使其贯穿老街巷，焕发新活力。为了保护完整的文物建筑，围绕建筑周边布置河流走向，并在河道两侧种植多种花卉及绿植。

This project is one of the historical and cultural reserves closest to Tian'anmen Square. The design, which was developed with advanced planning concepts, integrated the existing situation of the city and the river. Under the concept of "Bringing new life into old hutongs", we introduced the idea of "human living environment" and "urban rejuvenation", and we proposed eight goals: disaster reduction, livelihood improvement, cultural rebuilding, urban context darning, space improvement, cultural inheritance, ecological restoration and facilities enhancement. Retaining the original layout of hutongs and roads, we improved the living environment of residents and further the quality of their life. We also repaired the water system of the Sanli River in Qianmen, injecting new vitality as it runs through the old streets. In order to keep heritage building intact, we redirected the path of the river around the building creating a lush landscape on both sides.

附录
APPENDIX

文化 / 体育
SPORTS BUILDING

1 突尼斯青年之家
Tunisia Youth Hotel
建成时间 Completion Year: 1990
建设面积 Building Area: 7 000 m^2

2 新疆体育中心体育场
Xinjiang Sports Center Stadium
建成时间 Completion Year: 2005
建设面积 Building Area: 75 000 m^2

1 | 4 | 5

3 南通博物苑
Nantong Museum
建成时间 Completion Year: 2006
建设面积 Building Area: 6 393 m^2

4 太仓图博中心
Taicang Library and Museum Center
建成时间 Completion Year: 2010
建设面积 Building Area: 36 617 m^2

2 | 3 | 6

5 青岛国际帆船中心
Qingdao International Sailing Center
建成时间 Completion Year: 2006
建设面积 Building Area: 137 599 m^2
合作公司 Partner: 澳大利亚 COX 公司 COX ARCHITECTURE

6 成都安仁民国风情街
Chengdu Anren Minguo-Style Street
建成时间 Completion Year: 2012
建设面积 Building Area: 13 000 m^2

7
9
10

7 北京妫河建筑创意区接待中心

Beijing Gui River Architecture Innovation Area Reception Center

建成时间 Completion Year: 2013

建设面积 Building Area: 3 595 m^2

8 深圳市青少年活动中心

Shenzhen Youth Activity Center

建成时间 Completion Year: 2016

建设面积 Building Area: 38 171.5 m^2

9 长沙梅溪湖国际文化艺术中心

Changsha Meixi Lake International Culture Art Center

建成时间 Completion Year: 2017

建设面积 Building Area: 126 012 m^2

合作公司 Partner: 英国扎哈·哈迪德建筑事务所 Zaha Hadid Architects

10 峨眉武术文化小镇（峨眉印）

Emei Martial Arts Culture Town (Amazing)

建成时间 Completion Year: 2017

建设面积 Building Area: 75 545 m^2

11 北京市档案馆新馆

New Beijing Municipal Archives

建成时间 Completion Year: 2018

建设面积 Building Area: 114 988 m^2

12 张家口奥体中心

Zhangjiakou Olympic Center

设计时间 Design Began: 2015

建设面积 Building Area: 3 595 m^2

合作公司 Partner: 德国奥尔韦伯合伙人建筑设计有限公司 Auer Weber Architekten

8
11
12

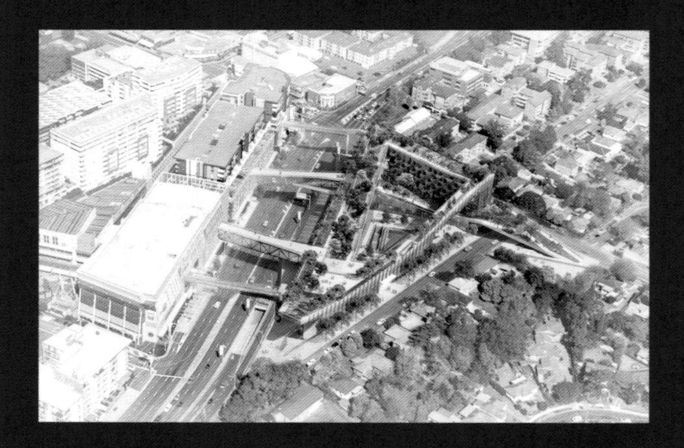

13 / 16 / 17

13 澳大利亚悉尼莱德市市民中心

Ryde City Hub Sydney Australia

设计时间 Design Began: 2016

建设面积 Building Area: 23 500 m^2

观演 / 博览
PERFORMANCE / EXHIBIT

14 内蒙古乌兰恰特大剧院、内蒙古博物馆

Ulaan Tiyatr Grand Theatre and Museum of Inner Mongolia

建成时间 Completion Year: 2007

建设面积 Building Area: 96 071.57 m^2

合作公司 Partner: 日本安井 GLAnet

15 中国（太原）煤炭交易中心

China Taiyuan Coal Transaction Center

建成时间 Completion Year: 2011

建设面积 Building Area: 114 600 m^2

16 重庆国际博览中心

Chongqing International Expo Center

建成时间 Completion Year: 2013

建设面积 Building Area: 602 189 m^2

合作公司 Partner: TDP

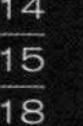

14 / 15 / 18

17 哈尔滨大剧院

Harbin Opera House

建成时间 Completion Year: 2015

建设面积 Building Area: 79 396 m^2

合作公司 Partner: MAD

18 又见敦煌剧场

Encore Dunhuang Theater

建成时间 Completion Year: 2016

建设面积 Building Area: 19 901 m^2

19 2016 年唐山世界园艺博览会景观规划设计

Landscape Planning and Design International Horticultural Exhibition 2016 Tangshan

建成时间 Completion Year: 2016

建设面积 Construction Area: 5.4km^2

20 福州数字中国会展中心

Fuzhou Digital China Convention and Exhibition Center

建成时间 Completion Year: 2019

建设面积 Building Area: 115 903.4 m^2

19 | 21 | 22

20 | 23 | 24

教育 / 科研
EDUCATIONAL / RESEARCH CENTER

21 北京第四中学

Beijing No.4 High School

建成时间 Completion Year: 1987

建设面积 Building Area: 17 400 m^2

22 北京顺义国际学校

International School of Beijing

建成时间 Completion Year: 2001

建设面积 Building Area: 69 000 m^2

23 北京大学国际关系学院（国政楼）

School of International Studies (Guozheng Building) Peking University

建成时间 Completion Year: 2004

建设面积 Building Area: 9 997 m^2

24 清华大学美术学院教学楼

Teaching Building for Academy of Arts & Design Tsinghua University

建成时间 Completion Year: 2005

建设面积 Building Area: 60 890 m^2

25
28
30

25 北京航空航天大学东南区教学科研楼

Southeast Teaching and Research Building of Beihang University

建成时间 Completion Year: 2006

建设面积 Building Area: 226 500 m^2

26 北京第二实验小学

Beijing No.2 Experimental Primary School

建成时间 Completion Year: 2007

建设面积 Building Area: 31 634 m^2

27 全国组织干部学院

National Organizations and Leaders School

建成时间 Completion Year: 2010

建设面积 Building Area: 36 700 m^2

28 深圳海上运动基地暨航海运动学校

Shenzhen Maritime Sports Base and Marine Navigation Sports School

建成时间 Completion Year: 2011

建设面积 Building Area: 27 180 m^2

26
27
29
31

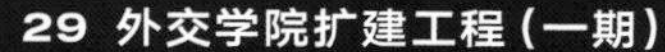

29 外交学院扩建工程（一期）

Extension Project of China Foreign Affairs University (Phase I)

建成时间 Completion Year: 2012

建设面积 Building Area: 80 958 m^2

30 北京爱慕内衣生产厂房

Aimer Underwear Factory Beijing

建成时间 Completion Year: 2013

建设面积 Building Area: 53 000 m^2

31 北京航空航天大学南区科技楼

South Technology Building of Beihang University

建成时间 Completion Year: 2015

建设面积 Building Area: 225 000 m^2

32
34
36

33
35
37

32 北京 161 中学回龙观学校

Beijing No. 161 High School
Huilongguan Branch School
建成时间 Completion Year: 2015
建设面积 Building Area: 225 000 m^2

33 腾讯广州总部

Tencent Guangzhou Headquarters
设计时间 Design Began: 2016
建设面积 Building Area: 150 000 m^2
合作公司 Partner: 让·努维尔建筑事务所
Ateliers Jean Nouvel

办公 / 商业
OFFICES / COMMERCIAL

34 国家海洋局

State Oceanic Administration
建成时间 Completion Year: 1980
建设面积 Building Area: 27 944 m^2

35 蛇口招商局办公楼

China Merchants Shekou Office Building
建成时间 Completion Year: 1982
建设面积 Building Area: 7 811 m^2

36 中国社会科学院办公楼

Office Building of Chinese Academy
of Social Sciences
建成时间 Completion Year: 1984
建设面积 Building Area: 43 667 m^2

37 台盟总部办公楼

Taiwan Democratic Self-government
League Headquarters
建成时间 Completion Year: 1984
建设面积 Building Area: 2 100 m^2

38 国家海关总署办公楼

Office Building of General Administration of Customs R.R. China

建成时间 Completion Year: 1990

建设面积 Building Area: 26 600 m^2

39 交通部办公楼

Office Building of Ministry of Transport of the People's Republic of China

建成时间 Completion Year: 1994

建设面积 Building Area: 47 000 m^2

40 建威大厦

Canway Building

建成时间 Completion Year: 1997

建设面积 Building Area: 60 232.78 m^2

41 北京图书大厦

Beijing Book Building

建成时间 Completion Year: 1997

建设面积 Building Area: 47 000 m^2

42 国际投资大厦

International Investment Building

建成时间 Completion Year: 2005

建设面积 Building Area: 147 096 m^2

43 凯晨广场

Chemsunny World Trade Center

建成时间 Completion Year: 2006

建设面积 Building Area: 194 206 m^2

合作公司 Partner: SOM 建筑设计事务所 Skidmore, Owings & Merrill LLP

38 / 41 / 43

39 / 40 / 42

44 北京国际图书城（北京出版发行物流中心）

Beijing International Book City (Beijing Publishing and Distribution Logistics Center)

建成时间 Completion Year: 2007

建设面积 Building Area: 76 300 m^2

45 中国工商银行业务营运中心

Business Operation Center of Industrial and Commercial Bank of China

建成时间 Completion Year: 2010

建设面积 Building Area: 65 500 m^2

合作公司 Partner: 贝氏建筑事务所

Pei Partnership Architects

45 / 47 / 49

46 国家开发银行

China Development Bank

建成时间 Completion Year: 2012

建设面积 Building Area: 149 570 m^2

合作公司 Partner: 中科院建筑设计研究院

Institute of Architecture Design and Research Academy of Sciences

47 银河 SOHO

Yinhe SOHO

建成时间 Completion Year: 2012

建设面积 Building Area: 330 117 m^2

合作公司 Partner: 英国扎哈·哈迪德建筑事务所

Zaha Hadid Architects

44 / 46 / 48

48 重庆大厦

Chongqing Building

建成时间 Completion Year: 2014

建设面积 Building Area: 108 441 m^2

49 中国驻澳大利亚新建大使馆

New Embassy of the People's Republic of China in the Commonwealth of Australia

建成时间 Completion Year: 2013

建设面积 Building Area: 6 858 m^2

50
52
53
54

50 三亚海棠湾国际购物中心一期

Sanya Haitang Bay International Shopping Center Phase I

建成时间 Completion Year: 2014

建设面积 Building Area: 119 360.92 m^2

合作公司 Partner: 法国 VP 建筑设计事务所 Valode & Pistre Architects

51 长沙北辰新河三角洲 A1 地块城市综合体

Urban Complex of Changsha Beichen Xinhe Delta Plot A1

建成时间 Completion Year: 2014

建设面积 Building Area: 318 739.25 m^2

合作公司 Partner: RTKL / JERDE

52 保利国际广场 1、3 号楼

Building No.1 and 3 of Poly International Plaza

建成时间 Completion Year: 2014

建设面积 Building Area: 167 974 m^2

合作公司 Partner: SOM 建筑设计事务所 Skidmore, Owings & Merrill LLP / 筑博设计股份有限公司 ZHUBO Design Co. Ltd.

53 重庆新光天地

Chongqing Shin Kong Place

建成时间 Completion Year: 2015

建设面积 Building Area: 264 625.41m^2

合作公司 Partner: Nikken Sekkei LTD / SKM Architects / SWA Group

54 上海嘉定大融城

Shanghai Jiading IMIX Park

建成时间 Completion Year: 2016

建设面积 Building Area: 126 439 m^2

55 丽泽 SOHO

Lize SOHO

设计时间 Design Began: 2014

建设面积 Building Area: 172 800m^2

合作公司 Partner: 英国扎哈·哈迪德建筑事务所 Zaha Hadid Architects

51
55

56 CBD 核心区 Z6 项目

Beijing CBD Z6

设计时间 Design Began: 2012

建设面积 Building Area: 245 000 m^2

酒店 / 会议
HOTEL / CONFERENCE

57 长富宫中心

Hotel New Otani Chang Fu Gong

建成时间 Completion Year: 1989

建设面积 Building Area: 95 000 m^2

58 北京国际会议中心

Beijing International Convention Center

建成时间 Completion Year: 1990

建设面积 Building Area: 45 000 m^2

59 大观园酒店

Grand View Garden Hotel

建成时间 Completion Year: 1993

建设面积 Building Area: 36 000 m^2

56
59
61

60 博鳌金海岸大酒店

Bo'ao Golden Coast Hot Spring Hotel

建成时间 Completion Year: 1997

建设面积 Building Area: 21 189 m^2

57
58
60

61 三亚凯宾斯基酒店

Kempinski Hotel Haitang Bay Sanya

建成时间 Completion Year: 2007

建设面积 Building Area: 59 176 m^2

合作公司 Partner: WATG

62
―
65

62 北京香港马会会所

Beijing Hong Kong Jockey Club Clubhouse

建成时间 Completion Year: 2009

建设面积 Building Area: 37 809 m^2

63 奥林匹克公园（B 区）会议中心配套设施

Convention Center Supporting Facilities of Olympic Park (B Area)

建成时间 Completion Year: 2009

建设面积 Building Area: 263 818 m^2

合作公司 Partner: 英国 RMJM 建筑设计事务所 RMJM / 美国 NBBJ 建筑设计公司 NBBJ

64 西双版纳避寒皇冠假日度假酒店（一期）

Crowne Plaza Resort Xishuangbanna Parkview (Phase I)

建成时间 Completion Year: 2012

建设面积 Building Area: 92 534.26m^2

合作公司 Partner: 洲际酒店集团 InterContinental Hotels Group / 泰国 P49 Deesign 室内设计公司 P49 Deesign / 新加坡博绿地安（亚洲）有限公司 Peridian Asia Pte Ltd. / 泰国 Habita 建筑师事务所 Habita Architects

65 北京华尔道夫酒店

Waldorf Astoria Beijing

建成时间 Completion Year: 2014

建设面积 Building Area: 44 180 m^2

合作公司 Partner: 美国 AS+GG 建筑设计师事务所 Adrian Smith + Gordon Gill Architecture

66 紫旸山庄

Ziyang Villa

建成时间 Completion Year: 2016

建设面积 Building Area: 40 000 m^2

63
―
64
―
66

67
69
71

交通 / 医疗
TRANSPORTATION / HEALTHCARE

67 北京医科大学附属人民医院病房楼
Ward Building of Peking University Health Science Center Affiliate People' s Hospital
建成时间 Completion Year: 1987
建设面积 Building Area: 23 405 m^2

68 协和医院业务楼

Practicing Building of Peking Union Medical College Hospital
建成时间 Completion Year: 1991
建设面积 Building Area: 59 800 m^2

69 北京同仁医院扩建工程
Extension Project of Beijing Tongren Hospital
建成时间 Completion Year: 1991
建设面积 Building Area: 46 555 m^2

70 北京西客站
Beijing West Railway Station
建成时间 Completion Year: 1995
建设面积 Building Area: 50 000 m^2

71 北京友谊医院综合楼
Complex Building of Beijing Friendship Hospital
建成时间 Completion Year: 2005
建设面积 Building Area: 53 000 m^2

72 银川火车站
Yinchuan Railway Station
建成时间 Completion Year: 2012
建设面积 Building Area: 30 000 m^2

68
70
72

73 南宁吴圩国际机场 T2 航站楼

Terminal 2 of Nanning Wuxu International Airport
建成时间 Completion Year: 2014
建设面积 Building Area: 189 000 m^2
合作公司 Partner: 上海民航新时代机场设计研究院有限公司 Shanghai Civil Aviation New Era Airport Design & Research Ltd. / KPF

74 北京地铁 6 号线通州核心区三站

Three Stations of Beijing Subway Line 6 in Tongzhou Central District
建成时间 Completion Year: 2016
建设面积 Building Area: 35 000 m^2

74
76
77

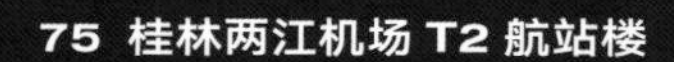

75 桂林两江机场 T2 航站楼

Terminal 2 of Guilin Liangjiang International Airport
建成时间 Completion Year: 2018
建设面积 Building Area: 100 000 m^2

76 前门交通一体化

Transportation Integration of Qianmen
设计时间 Design Began: 2014
建设面积 Building Area: 24 000 m^2

73
75
78

77 大连机场

Dalian Zhoushuizi International Airport
设计时间 Design Began: 2015
建设面积 Building Area: 491 000 m^2

78 海口机场

Haikou Meilan International Airport
设计时间 Design Began: 2015
建设面积 Building Area: 300 000 m^2

79
82
84

居住建筑
RESIDENTIAL ARCHITECTURE

79 北京恩济里小区

Beijing Enjili Housing Estate
建成时间 Completion Year: 1993
建设面积 Building Area: 136 200 m^2

80 浙江平湖梅兰苑规划及住宅

Planning and Residential Building of Meilan Garden Housing Estate Pinghu Zhejiang
建成时间 Completion Year: 2001
建设面积 Building Area: 82 000 m^2

81 北京星河湾小区一期

Beijing Xinghewan Housing Estate Phase I
建成时间 Completion Year: 2005
建设面积 Building Area: 248 800 m^2

82 北京世纪华侨城一期

Beijing Shiji Overseas Chinese Town Phase I
建成时间 Completion Year: 2006
建设面积 Building Area: 152 000 m^2

83 马连洼竹园住宅小区（西山华府）C 区住宅

C Area of Malianwa Bamboo Garden Housing Estate (Xishan Huafu)
建成时间 Completion Year: 2008
建设面积 Building Area: 104 320.72 m^2

84 嘉润园国际社区 C9 区

Jiarun Garden International Community C9
建成时间 Completion Year: 2009
建设面积 Building Area: 107 350.95 m^2

80
81
83

85
86
88

85 望京 A1 区 C 组团

Block C of Wang Jing A1 District
建成时间 Completion Year: 2009
建设面积 Building Area: 245 500 m²

86 银亿上上城

Yinyi Shangshang Town Housing Estate
建成时间 Completion Year: 2010
建设面积 Building Area: 218 154 m²

87 丰台区张仪村西城区旧城保护定向安置房

Old-city Preservation Oriented Affordable Housing of Xi Cheng District in Zhangyi Village Fengtai District
建成时间 Completion Year: 2014
建设面积 Building Area: 560 000 m²

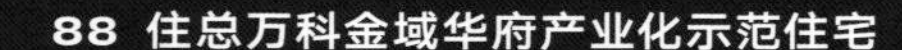

88 住总万科金域华府产业化示范住宅

Industrialization Housing Demonstration of Zhuzong Vanke Jinyu Huafu Housing Estate
建成时间 Completion Year: 2015
建设面积 Building Area: 11 838 m²

89 望京金茂悦住宅

Jin Mao Yue Housing Estate in Wangjing
建成时间 Completion Year: 2015
建设面积 Building Area: 122 961 m²

90 长安运河

Chang'an Canal
设计时间 Design Began: 2012
建设面积 Building Area: 100 000 m²
合作方 Partner: 非常建筑 Atelier FCJZ / 李玮珉建筑事务所 LWMA/ SWA / ARUP

87
89
90

91
93
94

改造 / 修复
RENOVATION / RESTORATION

91 北京市委办公楼修缮改造

Restoration of Office Building for Beijing Committee of the Communist Party of China

建成时间 Completion Year: 2014

建设面积 Building Area: 77 000 m^2

92 全国人大常委会会议厅改扩建

Renovation of Conference Hall for Executive Committee of the National People' s Congress of the People' s Republic of China

建成时间 Completion Year: 2012

建设面积 Building Area: 34 110 m^2

合作公司 Partner: 清华大学建筑学院 School of Architecture, Tsinghua University / 华堂建筑装修工程有限公司 Huatang Construction Decoration Engineering co. LTD

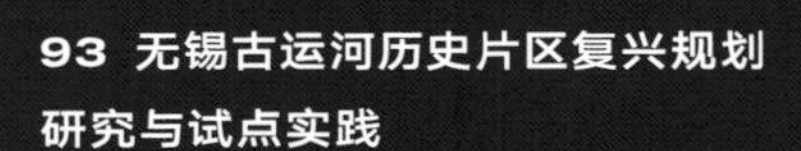

93 无锡古运河历史片区复兴规划研究与试点实践

Revitalization Planning Research and Demonstrative Practice of Wuxi Ancient Canal Historical Area

建成时间 Completion Year: 2011

规划面积 Planning area: 187 800 m^2

合作公司 Partner: 清华大学建筑学院 School of Architecture, Tsinghua University

92
95

94 新首钢高端产业综合服务区

Comprehensive Service Area of New Shougang High-end Industry

设计时间 Design Began: 2009

规划面积 Planning Area: 8.63km^2

95 三联韬奋书店三里屯店

SDX Joint Publishing Company Taofen Bookstore in Sanlitun

建成时间 Completion Year: 2018

建设面积 Building Area: 700 m^2

96
98
100

城市设计及其他
URBAN DESIGN & OTHERS

96 北京琉璃厂文化街

Beijing Colored-glaze Factory Cultural Street

建成时间 Completion Year: 1985

建设面积 Construction Area: 34 000m^2

97 西单文化广场

Xidan Cultural Plaza

建成时间 Completion Year: 1999

建设面积 Construction Area: 46 600m^2

98 长安街及延长线环境提升设计

Environmental Improvement Design of Chang An Avenue and Extension Area

建成时间 Completion Year: 2016

规划面积 Planning Area: 18.45km^2

97
99

99 北京新机场临空经济区

Beijing New Airport Economic Zone

设计时间 Design Began: 2017

规划面积 Planning Area: 50km^2

100 什刹海环湖绿道（前海）整体设计与提升

Integrated Design and Improvement of Shichahai Riverfront Green Avenue (Qianhai)

设计时间 Design Began: 2017

规划面积 Planning Area: 58 000m^2

谨以此书献给对北京市建筑设计研究院有限公司给予支持、关心和帮助的各界朋友们

We dedicate this book to all friends who have given support, concern and help to BIAD

项目索引
PROJECT INDEX

主题篇章
LANDMARK PROJECTS

壹 | 建国基业 · 天安门广场建筑群建设
TIAN'ANMEN SQUARE ARCHITECTURAL COMPLEX

14 天安门城楼重建工程 / 天安门观礼台
The Restoration Project of the Tian'anmen Square Rostrum / Tian'anmen Square Viewing Platform

15 毛主席纪念堂
Chairman Mao Memorial Hall

16 人民大会堂
Great Hall of the People

17 中国革命博物馆与中国历史博物馆
Chinese Revolution Museum and Chinese History Museum

贰 | 开放契机 · 北京亚运会场馆建设
BEIJING ASIAN GAMES VILLAGE

22 国家奥林匹克体育中心 一期
National Olympic Sports Center Phase I

24 北京工人体育场
Beijing Workers'Stadium

25 第十一届亚运会运动员村及配套设施
The 11th Asian Games Village and Supporting Facilities

叁 | 入世宣言 · 北京 CBD 核心区建设
THE BEIJING CBD CORE AREA

30 中信大厦
CITIC Tower

32 北京 CBD 核心区地下公共空间市政交通基础设施及配套工程
Municipal Transportation Infrastructure Project in Underground Public Space of the Beijing CBD Core Area

33 中国人寿金融中心（北京）
China Life Financial Center (Beijing)

肆 | 联通世界 · 首都的机场建设
AIRPORT CONSTRUCTION IN THE CAPITAL

38 首都国际机场 T1 航站楼
Terminal 1 of Beijing Capital International Airport

38 首都国际机场 T2 航站楼
Terminal 2 of Beijing Capital International Airport

39 首都国际机场 T3 航站楼
Terminal 3 of Beijing Capital International Airport

40 北京大兴国际机场
Beijing Daxing International Airport

伍 | 崛起之路 · 北京奥运建设
BEIJING OLYMPIC GAMES

46 北京奥林匹克公园中心区景观设计
The Landscape Design of the Beijing Olympic Park Central District

47 国家体育馆
National Stadium

48 五棵松体育馆
Wukesong Stadium

49 国家会议中心
National Conference Center

50 北京冬奥村人才公租房
Beijing Winter Olympic Village Public Housing for Athletes

51 国家速滑馆
National Speed Skating Oval

51 国家游泳中心改造
Reconstruction of the National Swimming Center

陆 | 盛世华章 · 主场外交建设
HOME COURT DIPLOMACY COMPLEX

55 博鳌亚洲论坛会议中心暨索菲特大酒店
Bo'ao Forum for Asia Conference Center

56 博鳌国宾馆
Bo'ao State Guesthouse

56 博鳌亚洲论坛永久会址二期
Bo'ao Forum for Asia Permanent Venue, Phase II

57 雁栖湖会议中心：第 22 届 APEC 峰会主会场 / 一带一路国际合作高峰论坛主会场
Yanqi Lake Conference Center (Main Venue of the 22th APEC Summit / Main Venue of the Belt and Road Forum for International Cooperation)

58 杭州国际博览中心改造：第 11 届 G20 峰会主会场
Hangzhou International Expo Center Renovation (Main Venue of the 11th G20 Summit)

59 厦门国际会议中心改造：第 9 届金砖会晤主会场
Renovation of the Xiamen International Conference Center (Main Venue of the 9th BRIC Meeting)

61 北京雁栖湖生态发展示范区规划景观综合提升设计
Integrated Improvement Design of Beijing Yanqi Lake Ecological Development Demonstration Area Planning Landscape

柒 | 走向世界 · 海外工程建设
OVERSEA CONSTRUCTION

63 中国驻英国使馆新馆舍改扩建工程（实施方案）
The Extension of Embassy of the People's Republic of China in United Kingdom (Implementing Plan)

64 明斯克北京饭店
Beijing Hotel Minsk

64 卡塔尔卢塞尔体育场钢结构与屋顶索膜结构工程
Steel structure and roof cable membrane structure of the Qatar Lusail Stadium

65 斯里兰卡国家艺术剧院
Sri Lanka National Art Theatre

65 援塔吉克斯坦议会大楼
Tajikistan Parliament Building

66 多哥共和国总统府
Presidential Palace of the Republic of Togo

67 缅甸国际会议中心
Myanmar International Convention Center

69 白俄罗斯标准游泳馆
Republic of Belarus Standard Swimming Pool

捌 | 和谐共生 · 城市环境设计
INTERNATIONAL GARDEN EXPO

72 2019 年北京世界园艺博览会国际馆
The International Pavilion of the Beijing World Horticultural Expo 2019

74 北京植物园展览温室
Beijing Botanical Park Greenhouse

75 中国园林博物馆
Chinese Garden Museum

77 2016 年唐山世界园艺博览会
2016 Tangshan World Horticultural Exposition

玖 | 北京故事 · 首都城市更新改造
CAPITAL URBAN REGENERATION

79 北京坊城市更新
Urban Regeneration of Beijing Fun

80 全国妇联办公楼改扩建工程
Renovation and Expansion of the National Women’s Federation Office

81 华电（北京）热电有限公司天宁寺厂区
Huadian (Beijing) Thermal Power Plant in Tianning Temple

82 北京什刹海鼓楼西大街复兴工程
The Renovation and Revival Plan for West Gulou Street in Beijing

83 王府井街道整治之口袋公园
Renovaiton of the Pocket Park of Wangfujing Street

83 北京西打磨厂 222 号院
Beijing West Grinding Factory NO.222

85 前门东区城市更新
Qianmen East Urban Regeneration

拾 | 面向未来 · 设计的创新与突破
DESIGN INNOVATION AND BREAKTHROUGH

87 丽泽 SOHO
Lize SOHO

88 凤凰中心
Phoenix Center

90 500m 口径球面射电望远镜结构工程
The 500-meter Aperture Spherical Radio Telescope (FAST)

90 江门中微子实验中心探测器主体结构
The Jiangmen Underground Neutrino Observatory (JUNO) Center, Stainless-steel Main Structure of Detector

公共建筑
PUBLIC BUILDING

文化 / 体育建筑
CULTURE / SPORTS BUILDING

96 深圳文化中心
Shenzhen Cultural Center

100 深圳湾体育中心
Shenzhen Bay Sports Center

104 故宫北院
North Annex of the Palace Museum

106 首都图书馆新馆
New Extension of the Capital Library (Beijing)

108 炎黄艺术馆
Yanhuang Art Meseum

110 国家美术馆
The National Gallery of China

112 嘉德艺术中心
Jiade Art Center

观演 / 博览建筑
PERFORMANCE / EXPO BUILDING

114 中国国际展览中心 2 ~ 5 号馆
Hall 2–5 of the China International Exhibition Center

118 又见五台山剧院
Encore Mount Wutai

122 北京动物园大熊猫馆
The Panda House of the Beijing Zoo

124 珠海大剧院
Zhuhai Opera House

126 中国科学技术馆
China Science and Technology Museum

128 中国电影博物馆
China National Film Museum

130 中国国际展览中心 I 期
China International Exhibition Center Phase I

132 国家大剧院
The National Center for the Performing Arts of China

134 第九届中国（北京）国际园林博览会主展馆
Main Exhibition Hall of the 9th China (Beijing) International Garden Expo

教育 / 科研建筑
EDUCATION / RESEARCH BUILDING

136 中国美术学院南山校区
Nanshan Campus of China Academy of Art

140 北京建筑大学 6 号综合服务楼
Service Building No. 6 of the Beijing University of Architecture

144 北川中学
Beichuan Middle School

148 望京科技园二期
Wangjing Science and Technology Park Phase II

150 联想园区 C 座
Block C of Lenovo Park

152 联想研发基地
Lenovo Research and Development Base

154 腾讯北京总部
Tencent Beijing Headquarters

156 小米科技北京总部
Xiaomi Technology Headquarters Beijing

158 低碳能源研究所及神华技术创新基地
Low Carbon Energy Research Institute and Shenhua Technology Innovation Base

160 国电新能源技术研究院
Guodian New Energy Technology Research Institute

办公 / 商业建筑
OFFICE / COMMERCIAL BUILDING

162 中国石油大厦
China Petroleum Building

166 侨福芳草地
Parkview Green

170 北京城市副中心行政办公区
Administrative Offices of the Beijing City Sub-Center

174 北京奥体商务南区 OS-10B 城奥大厦
South Olympic Business District OS-10B Cheng'ao Building

178 全国人大机关办公楼
Administrative Office of the National People's Congress(NPC)

180 商务部办公楼改造
Ministry of Commerce Office Building Renovation

182 北京国际金融大厦
Beijing International Finance Building

184 北京市高级人民法院
Beijing Higher People's Court

186 恒基中心
Henderson Center

188 北京电视中心
Beijing TV Center

190 王府中环
Wangfu Central

192 蓝色港湾
Blue Harbor

194 法国驻华大使馆新馆
New Embassy of France in China

酒店 / 会议建筑 HOTEL / CONFERENCE BUILDING

196 三亚太阳湾柏悦酒店
Park Hyatt Sanya Sunny Bay Resort

198 昆仑饭店
The Kunlun Hotel

200 钓鱼台国宾馆会议中心
Conference Center of Diaoyutai State Guesthouse

医疗 / 交通建筑 MEDICAL / TRANSPORTATION BUILDING

202 昆明长水国际机场航站楼
Kunming Changshui International Airport Terminal Building

206 广州南站
Guangzhou South Railway Station

210 首都医科大学附属北京天坛医院
Beijing Tiantan Hospital, Capital Medical University

214 深圳宝安国际机场 T3 航站楼
Shenzhen Baoan International Airport Terminal 3

216 青岛北站
Qingdao North Railway Station

218 南京南站主站房
Main Station of the Nanjing South Railway Station

220 北京儿童医院门诊楼
Beijing Children's Hospital Outpatient Building

222 首都医科大学宣武医院改扩建一期工程
Xuanwu Hospital Capital Medical University Reconstruction and Extension Phase I

居住建筑 RESIDENTIAL BUILDING

226 万科第五园
The Fifth Vanke Village

230 北京 SOHO 现代城
SOHO New Town Beijing

232 万科中粮假日风景（万恒家园二期）D 地块 D1、D8 工业化住宅
Parcel D1 and D8 Industrialized Housing of Vanke COFCO Holiday Town (Wanheng Homeworld Phase Ⅱ)

234 建国门外外交公寓
Jianwai Diplomatic Residence Compound

城市设计及其他 URBAN DESIGN AND OTHERS

238 丽泽金融商务区城市设计
Urban Design of Lize Financial Business District

242 怀柔科学城城市设计
Huairou Science City

246 西安唐大明宫国家大遗址保护展示示范园区
Xi'an Daming Palace National Heritage Protection and Demonstration Area

250 前门东三里河复兴
Qianmen Dongsanli River Regeneration

图书在版编目（CIP）数据

北京市建筑设计研究院有限公司作品集：1949-2019/《北京市建筑设计研究院有限公司作品集：1949-2019》编委会编. -- 上海：同济大学出版社，2020.1
（北京市建筑设计研究院有限公司（BIAD）成立70周年（1949-2019）系列丛书）
ISBN 978-7-5608-8877-4

Ⅰ. ①北… Ⅱ. ①北… Ⅲ. ①建筑设计－作品集－中国－现代 Ⅳ. ① TU206

中国版本图书馆CIP数据核字(2019)第271525号

北京市建筑设计研究院有限公司作品集1949－2019

《北京市建筑设计研究院有限公司作品集1949－2019》编委会 编

出 版 人：华春荣
策　　划：群岛工作室
责任编辑：李争
责任校对：徐春莲
装帧设计：Plankton Design
版　　次：2020年1月第1版
印　　次：2020年1月第1次印刷
印　　刷：联城印刷（北京）有限公司
开　　本：787mm×1092mm　1/12
印　　张：23
字　　数：580 000
书　　号：ISBN 978-7-5608-8877-4
定　　价：288.00元
出版发行：同济大学出版社
地　　址：上海市杨浦区四平路1239号
邮政编码：200092
网　　址：http://www.tongjipress.com.cn
经　　销：全国各地新华书店
本书若有印装质量问题，请向本社发行部调换。

光明城联系方式：info@luminocity.cn

BIAD SELECTED WORKS 1949－2019

Edited by: *Beijing Institute of Architectural Design Selected Works 1949－2019* Editorial Committee

ISBN 978-7-5608-8877-4
Initiated by: Studio Archipelago
Produced by: Hua Chunrong (publisher), Li Zheng (editing), Xu Chunlian (proofreading), Plankton Design (graphic design)
Published in January 2020, by Tongji University Press, 1239, Siping Road, Shanghai, China, 200092.
www.tongjipress.com.cn

Contact us: info@luminocity.cn